AF856138

FOUCAULT'S STRATA AND FIELDS

SYNTHESE LIBRARY

STUDIES IN EPISTEMOLOGY,

LOGIC, METHODOLOGY, AND PHILOSOPHY OF SCIENCE

Managing Editor:

JAAKKO HINTIKKA, *Boston University*

Editors:

DONALD DAVIDSON, *University of California, Berkeley*
GABRIËL NUCHELMANS, *University of Leyden*
WESLEY C. SALMON, *University of Pittsburgh*

VOLUME 218

MARTIN KUSCH
University of Oulu, Finland

FOUCAULT'S STRATA AND FIELDS

An Investigation into Archaeological and Genealogical Science Studies

SPRINGER SCIENCE+BUSINESS MEDIA, B.V.

Library of Congress Cataloging-in-Publication Data

Kusch, Martin.
Foucault's strata and fields : an investigation into archaeological and genealogical science studies / by Martin Kusch.
p. cm. -- (Synthese library ; v. 218)
Includes bibliographical references and indexes.
ISBN 978-94-010-5567-3 ISBN 978-94-011-3540-5 (eBook)
DOI 10.1007/978-94-011-3540-5
1. Foucault, Michel. 2. Science--Philosophy--History--20th century. 3. Social sciences--Philosophy--History--20th century.
I. Title. II. Series.
B2430.F724K87 1991
194--dc20 91-33950

ISBN 978-94-010-5567-3

Printed on acid-free paper

All Rights Reserved
© 1991 Springer Science+Business Media Dordrecht
Originally published by Kluwer Academic Publishers in 1991
Softcover reprint of the hardcover 1st edition 1991
No part of the material protected by this copyright notice may be reproduced or utilized in any form or by any means, electronic or mechanical, including photocopying, recording or by any information storage and retrieval system, without written permission from the copyright owner.

CONTENTS

ACKNOWLEDGEMENTS

I am indebted, most of all, to my colleagues and students in the Department of History of the University of Oulu. I have especially profited from discussions with Jouko Aho, Heini Hakosalo, Hippo Johansson, Jouko Jokisalo, Väinö Louekari, Matti Lukkarila, Juha Manninen, Aarne Nikkonen, Erkki Routti, Arja-Liisa Räisänen, Minna Uimonen, and Erkki Urpilainen. I am also grateful to "my" librarians, both Paula Viander of the Department of History and the people of the interlibrary loan office of the University Library. Outside of my alma mater, special thanks are due to Timo Airaksinen, Harri Kantele, Esa Itkonen, Karin Knorr-Cetina, Lothar Läsker, Tuomas Nevanlinna, Ilkka Niiniluoto, Calvin Normore, Kari Palonen, Outi Pasanen, Tuija Pulkkinen, Yves Roussel, Kari Saastamoinen, Hartmut Schröder, and Georg Henrik von Wright for their criticism and/or encouragement. My wife, Riitta Korhonen-Kusch, deserves special thanks not only for her commenting, but also for her crucial support at a point when I was ready to give up on the whole project. I am also grateful to an anonymous reviewer for Kluwer Academic Publishers.

The University of Oulu has supported my project with a travel-grant to Paris. The last stages of the work I carried out while holding the position of a younger fellow at the Academy of Finland, and while enjoying the hospitality of the University of Leuven (Belgium).

Finally, I wish to thank Ian Morris-Wilson and Jan Jagiellowicz for help with my English, Auli Kaipainen for excellent secretarial assistance, Jaakko Hintikka for accepting the book into Synthese Library, and Annie Kuipers of Kluwer Academic Publishers for her friendly cooperation.

I gratefully acknowledge the permissions to quote from the following books: Michel Foucault, *The Archaeology of Knowledge*, translated by A.M. Sheridan, Tavistock, London, 1974, and Michel Foucault, *Power/Knowledge: Selected Interviews & Other Writings 1972-1977*, copyright (c) 1972, 1975, 1976, 1977 by Michel Foucault, edited by C. Gordon, Pantheon Books, Random House, Inc., New York, 1980.

I dedicate this book to two friends who have had a large share in its creation and completion: Heini Hakosalo and Marja-Liisa Kakkuri-Knuuttila. Marja-Liisa Kakkuri-Knuuttila has been a close friend and supporter for many years now, and during the writing of this book we have spent many hours on the phone and in Chinese restaurants in Helsinki discussing Aristotle, rhetoric, feminism, vegetarianism, music and what not. As concerns the topic of this specific study, even more important has been Heini Hakosalo's cooperation. Since we set out to work on Foucault roughly around the same time, we had planned to publish our results as one book. As both of our parts kept growing in size, however, this plan had to be abandoned. Heini Hakosalo's part (Hakosalo 1991) therefore had to be published separately. Being my nextdoor neighbour in the history department, Heini Hakosalo has been not only a central source of inspiration and good humour, but also the first hurdle and final stumbling block for most of my ideas on Foucault.

Oulu, June 1991
Martin Kusch

INTRODUCTION

In recent years, a large number of books and articles on Foucault has been published. Almost all of the book-size studies are expository and introductory. Indeed, there seems to be no other modern philosopher with reference to whom a comparable number of introductions have been produced in such a short period. Most of the articles too provide overviews, rather than critical assessments or rational reconstructions, even though there exists by now a small number of fine papers also in the two latter genres. Moreover, more often than not, writers on Foucault approach his work as part and parcel of so-called "postmodern" philosophy. They concentrate on topics like the "death of the subject", the relation of Foucault's work to Derrida or Habermas, or its significance for postmodern art and culture.

Without wanting to deny the merits, either of introductory expositions, or of studies that read Foucault as a postmodern thinker, it seems to me that these received perspectives have tended to leave central areas and aspects of Foucault's work somewhat underexposed. As I see it, the most important of these areas are such as would suggest reading Foucault from the vantage point of recent developments in the philosophy, sociology and history of science.

During the last two decades or so, a number of new approaches have emerged and taken shape within philosophical, sociological and historical science studies. These new endeavors run under such catchword-labels as "anthropological" science studies, the "strong programme" of the sociology of science, "constructivism", "new experimentalism", "the reticulated model of scientific rationality", and "the feminist critique of science". While these approaches differ in many respects from one another, it still seems possible to point out two main features which many of them share.

First, a number of philosophers and historians have – at last – moved beyond the fruitless dispute over continuity and discontinuity in the history of science. The predominant question is no longer whether a scientific field develops either cumulatively or through total breaks. Instead, the crucial problem now is how to distinguish, within a scientific field of study, different forms of changes within different strata or layers of theoretical assumptions, problems, values, methods and techniques within a scientific culture.

Second, sociologists and feminists have begun to study scientific communities as "tribes", "cultures" and "power networks". For many writers in the sociology of science as well as for many feminist critics, science is no longer the purely cognitive, socially disinterested and gender-neutral, rational enterprise as which it was often presented by traditional philosophies of science. Rather, in a number of investigations published during the seventies and eighties, science is looked upon as one social culture (or bundle of cultures) among others, a culture whose beliefs, relations and actions can be studied with methods known from sociology and anthropology.

Now, to claim that Foucault can be fruitfully studied as contributing to these recent developments, is no radically new and contested thesis. The present study merely takes this thesis and tries to develop it further. That is to say, in this book, I present a reconstruction of those parts of Michel Foucault's writings that I feel are of considerable interest for both of these two main concerns in contemporary science studies. Going beyond a mere mechanical comparison, I shall propose various ways in which these more recent lines of thought in science studies, and a reconstructed, conceptually sharpened and, with respect to several key theses, newly defended Foucauldian project, can be used to illuminate and strengthen one another.

As concerns the move beyond continuity and discontinuity, Foucault is, according to the received view, a historian or philosopher who works on the assumption of radical ruptures in the history of science. However, while this conception might perhaps be partially justified as far as some of Foucault's less successful historical investigations are concerned, it certainly does not hold for his *discours de la méthode*, *The Archaeology of Knowledge*. In the first part of this study, I shall show that *The Archaeology of Knowledge* contains arguments against both one-sided continuist and excessive discontinuist models of scientific change. I shall also point out that *The Archaeology* includes a detailed discussion of various approaches to the history of science, ranging from the French *Annales* School and the historical epistemology of Bachelard and Canguilhem to the History of Ideas. Most importantly, the work outlines a theory of historical constructivism, a theory that is informed by the methodological writings of *Annales* historians.

In the second part of this study, I shall turn to the central concepts and ideas of Foucault's "genealogical" works. In these Nietzsche-inspired studies, written roughly between 1970 and his death in 1984, Foucault focusses more directly than in his earlier, archaeological phase on phenomena of power, especially the relations between scientific knowledge and social power. As I shall show, Foucauldian genealogy includes several of those theses that can roughly be grouped around the second main ingredient of the new science studies, to wit, the study of science as a social field. Foucault proposes a theory of social power and its relations to scientific knowledge as well as detailed methodological suggestions on how these interrelations might be studied from a historical point of view.

As should be clear from the above, my central interest in this study is topical, rather than historical or biographical. In other words, my overriding concern is not to determine what Foucault really meant or said - even though I obviously cannot ignore such questions completely - but rather to use his ideas for the purpose of contributing to the debates in the philosophy and sociology of science. In adopting this attitude towards Foucault's texts, I actually abide by his own view on how best to approach his writings. Commenting on his relation to Nietzsche, Foucault once wrote that "the only valid tribute to thought such as Nietzsche's is precisely to use it, to deform it, ... And if commentators then say that I am being faithful or unfaithful to Nietzsche, that is of absolutely no interest" (P/K 54).

Finally, a word is due on the relation between the two main parts of this study. I follow the received wisdom of Foucault research and divide his oeuvre into two phases. The first comprises Foucault's "archaeological" work, issuing in the afore-mentioned *discours de la méthode*, *The Archaeology of Knowledge*, published in 1969. The second phase is that of "genealogy", i.e. Foucault's writings from 1970 onwards. Of course, within both phases further distinctions could easily be made. However, since I am not primarily interested in Foucault's intellectual biography, questions of development will be discussed only occasionally, and only as far as the major changes are concerned.

I have not tried to make my reconstructions of Foucauldian archaeology and genealogy part of one overall, unified argument. Consequently, both parts can also be read separately. This is not to say, that the archaeological and the genealogical projects are without links and interrelations. To the contrary, already the key notion of Foucauldian archaeology, the notion of "discursive formation", refers to social conditions of science, and in the genealogical writings, Foucault spells out more clearly many of the philosophical tenets at work in his archaeology. Moreover, the critical thrust of *The Archaeology of Knowledge*, its criticism of anachronisms and taken for granted units like "discipline", "author", or "oeuvre", carries over into the genealogical edifice. Indeed, Foucault could have left behind the general historiographical proposals of *The Archaeology* only by paying the high price of falling back to traditional naive history. However, all of these interrelations notwithstanding, and despite my reluctance to follow those interpreters who construct something of a "break" between archaeology and genealogy, I have thought it best to discuss the two projects separately. Two reasons have been pivotal. First, the terminology within the two phases is rather different; and even though I did not intend to put forward merely another introductory book on Foucault, I still want to provide the reader with some help in finding her way around the Foucauldian conceptual jungle. Second, and more importantly, it seems to me that a separation of the two projects is called for by the compartmentalization of present day science studies themselves. Not everyone who is interested in the broad questions of continuity and discontinuity in the history of science, is likewise convinced, that the relation between scientific knowledge and social power is a question worth pondering upon, and not everyone who shares the genealogical interest in identifying the effects of exercises of power upon the human body, is likewise interested in detailed considerations of the relations betweens scientific statements of different disciplines. To bind together all of Foucault's suggestions into one single bundle, I fear, would have lessened the accessibility and the convincingness of what I take to be the most important tenets of his work; it would have made it difficult for me to argue that Foucault's theoretical work is not a diffuse tangle of ideas, but rather a clearly delimited set of proposals that are separately open to critical assessment.

PART I:

FOUCAULDIAN ARCHAEOLOGY

1. INTRODUCTION

Foucault's archaeological *discours de la méthode*, *The Archaeology of Knowledge* (AK) is much less widely read than all of his other books. In part this is due to the fact that this methodological treatise does not contain the kind of grand historical – speculative or sweeping – theorizing that readers have come to expect from this writer. The comparatively 'sober' nature of the book has not helped it much, however, in reaching the audience of specialists for whom it was written, historians in general and historians and philosophers of science in particular. One reason is that the majority of historians and philosophers – in France as well as elsewhere – have formed their predominantly negative perception of Foucauldian historiography on the basis of *The Order of Things* (OT). This work, published three years before *The Archaeology*, hardly lives up to strict standards of historical scholarship, and thus it is small wonder that practicing historians and methodologists do not feel any urge to study Foucault's methodological ideas.

Another reason for the small amount of attention that the book has received, is undoubtedly its difficulty.[1] Its intricacies are partly due to the fact that Foucault outlines his methodological suggestions by and large *via negationis* – a fact that has led Maurice Blanchot to speak of a "negative theology" (1987: 74), and Jean-François Revel to propose "Neither ... nor" as a more appropriate title (O'Farrell 1989: 15). The book is also hard to follow because it is written for an audience familiar not only with twentieth century French historiography (i.e. with various approaches within the *Annales* school, with the history of science of Duhem and Koyré, with the epistemology of Cavaillès, Bachelard and Canguilhem, the structuralist marxism of Althusser), but also with the history of ideas as represented by e.g., Lovejoy, Cassirer and Goldmann. Since general history and the history of science traditionally live separate lives, presupposing acquaintance with both cannot but further restrict the circle of potential readers.

Unfortunately, existing book-size presentations of Foucault reflect rather than remove these difficulties. While one author, Clare O'Farrell, simply bypasses the book with the disarming confession that *The Archaeology* "is a compelling book" only for those "who appreciate intricate formal geometrical structures in thought and method" (1989: 14), other interpreters see *The Archaeology* as something of an appendix to *The Order of Things*, as a series of corrections with respect to Foucault's earlier historical studies, or as an anticipation of his genealogical writings.

The first interpretative strategy is neatly exemplified not only by Gary Gutting, who writes that the "leitmotif of AK is ... the death of man" (1989: 227), but also by the best known commentary on Foucault's 'oeuvre', Hubert Dreyfus's and Paul Rabinow's *Michel Foucault. Beyond Structuralism and Hermeneutics* (1983). 'Bracketing' the methodological-historiographical dimension of *The Archaeology of Knowledge*, Dreyfus and Rabinow read the book as a systematic theory of language, compare it with Husserl, Heidegger, Kuhn and Searle, and ultimately fault Foucault for not having provided the "phenomenology to end all phenomenologies" which, as the authors see it, the book sets out to (or should?) develop (1983: 44–103).

The second tendency, i.e. reading *The Archaeology* as a mere series of afterthoughts on Foucault's earlier work, surfaces for instance in Gilles Deleuze or Charles Lemert and Garth Gillan. For Deleuze "[...] in this archaeology, Foucault [perhaps] offers us less a discourse on his method than the poem of his previous works, and reaches the point where philosophy is necessarily poetry, the severe poetry of what is said, which subsumes both nonsense and the greatest profundities" (1988: 18). Lemert and Gillan draw the conclusion that since *The Archaeology* is but a bunch of after-thoughts, they need to "disabuse readers of the idea that his *Archaeology of Knowledge* is of special importance. *Archaeology of Knowledge* is not a methodological statement ..." (1982: ix–x).

Finally, and to come to the third strategy, most interpreters struggle with the problem of how to situate *The Archaeology* within Foucault's intellectual biography. In the seventies Foucault adopts a different label, "genealogy", for his historical studies, and drops the term "archaeology". Commentators take this as a challenge to address the question whether Foucault is an "archaeologist of ideas or [a] genealogist of power?" (Smart 1985: 47). The usual answer is that – in part because of the events of May 1968 – Foucault turns Nietzschean, overcomes "the illusion of autonomous discourse" and his earlier role of "the detached spectator", and starts doing what he should have been doing, or intended to do, all along: study the relation between power and discourse (Dreyfus & Rabinow 1983: 103).

Here I do not wish to criticize these various interpretations in any detail, even though it is hard to suppress the remark that these studies could easily be used as illustrations for all of the categories of interpretatory excesses which *The Archaeology of Knowledge* enumerates. Instead, suffice it here to outline briefly my own concerns and preoccupations.

My reconstruction of *The Archaeology* is motivated by the view that this work contains a number of suggestions and criticisms concerning the writing of the history of science that are – or should be – of interest to present-day philosophers, sociologists and historians of science alike. To put it in a nutshell, *The Archaeology of Knowledge* merits such attention because its proposals and discussions link up with many of the topics that researchers in science studies have come to address over the last decade or so. To mention only a few catchwords, such topics and new interests are discourse analysis, research mentalities, styles of reasoning, the anthropology of science, different strata of assumptions involved in theorizing and experimenting, the relation between the history of ideas and the history of science, the historical links and structural similarities between disciplines, conceptualizations of science as a field of struggle over the modality of statements, and a new attitude with respect to the notorious problem of revolutions in science: the central question is no longer whether science is continuous or not, but rather which elements do and which do not accumulate.

Many of these new concerns have arisen from a heightened awareness of the anachronistic dangers in writing the history of science. This awareness, in turn, has been triggered, or has at least been reinforced, by a loosening of the borders between the history of science and other domains of historical research. Given this development, Foucault's significance need no longer come as a surprise. After all, Foucault was intimately familiar not only with the achievements of historians and philosophers of science like Bachelard and Canguilhem, but he was equally at home in the substantial historical as well as the methodological labors of the most important school in historical scholarship of this century, the *Annales* historians. Once we add the further observation that, already in the thirties, Bachelard formulated views on the development of science and the role of experimentation, more recent counterparts of which have sometimes been hailed as novelties in the sixties and the eighties respectively, it should be even less unexpected that a French text of 1969 can be of interest to scholars in the nineties.

It follows from the above-said that my reconstruction of *The Archaeology* needs to be "historical" as well as "rational". The reconstruction will be historical not in that it seeks to place this book within Foucault's intellectual biography or that it wants to present the book's ideas as successors to anticipations in the writings of others. Rather, the focus of the following reconstruction must be historical in so far as it attends not only to those views that, according to a somewhat unsophisticated categorization, are "Foucault's own", but also to ideas and theories "by others", i.e. ideas and theories which the book endorses. On the other hand, the reconstruction will be and must be "rational" in order for me to succeed in my attempt to convince the skeptical reader of the book's interest, that is, in order to make the book – or rather my reconstruction of it – a contribution to present concerns in the philosophy, history and sociology of science. Thus I do not abide by the view that "Foucault is not as opaque as Joyce, but like Joyce his style and his thought are inseparable" (Lemert & Gillan 1982: xii). Instead, I seek to formulate what I take to be the central theses of the book in a language that is as clear and precise as the inherent difficulty of the very topic allows for. Rational reconstruction also implies a criterion of selection. I do not attempt to provide a full commentary on all aspects of the book, nor do I engage the book in a debate over those concepts and ideas that are less important for my present concerns or that I regard as obscure; to add yet another deconstruction of the notoriously opaque notions like episteme or archive to the criticisms already found in the literature seems superfluous.

To conclude this introduction with the setting of some signposts, I shall start with a brief comment on the notion of archaeology, its use as a metaphor by other writers in France, and its employment by Foucault. Here I shall also make a brief suggestion on how to understand this metaphor. Subsequently, I shall review the exemplars of *Annalist* and historical epistemology, focussing especially on notions like mentality, discontinuity, series and time. This chapter will be followed by an analysis of Foucault's criticism of various anachronistic tendencies in the history of ideas and the history of science. As will be shown by way of examples, this criticism has not become obsolete. Finally, in the last two chapters, I shall reconstruct the "archaeological model" for studying the history of science. Here I shall compare the archaeology of knowledge with theories proposed more recently in the philosophy of science. My main point will be to argue that Foucauldian archaeology includes a model of scientific change that is not only a far cry from the excesses of *The Order of Things*, but also of some interest for present-day concerns.

2. ON THE VERY NOTION OF 'ARCHAEOLOGY'

In choosing the term 'archaeology' as a label for his approach to the history of science, Foucault took advantage of the fact that this term had already gained currency in France. Since the fifties, the concept of 'archaeology' (as well as the concept of 'geology') had been used to characterize the various fashionable intellectual movements of the day, movements like phenomenology, structuralist anthropology, psychoanalysis, Bachelardian epistemology, and *Annalist* history.

For instance, in his widely-read *Tristes Tropiques*, first published in 1955, Claude Lévi-Strauss had written of his "intense interest [in] geology". Geology allegedly allows us to pass beyond "the vast chaos" and "the various accidents of history and prehistory" in that it ultimately provides us with "the most majestic meaning of all ... [a meaning] that precedes, commands, and, to a large extent, explains the others". Lévi-Strauss had also explained his interest in anthropology, marxism and psychoanalysis by referring to their common geological dimension. In all four cases, what initially appear to be "impenetrable phenomena", or "a seemingly incoherent mass", turn out to have an "order" which is "neither contingent nor arbitrary"; in all these instances, "understanding consists in reducing one type of reality to another", where this reduction is based upon "a kind of *superrationalism*" that "will integrate the first [meaning] with the second, without sacrificing any of its properties" (1976: 69–71).

Psychoanalysis was also regarded as an essential archaeological enterprise by such renowned figures as Maurice Merleau-Ponty and Paul Ricoeur, both of whom couched the parallel between Husserlian phenomenology and psychoanalysis precisely in terms of the joint "archaeological" features of these two theories.[2] To speak of phenomenology as an archaeology had been common in France at least since Jean Cavaillès's *Sur la logique et la théorie de la science*, a work published in 1947.[3] Cavaillès, perhaps best known as a historian of mathematics, was one of the central figures in the early stages of French historical epistemology, a school of thought whose best known representatives are Gaston Bachelard and Georges Canguilhem. In 1959 Canguilhem characterized Bachelard's writings as amounting to an "archaeology of science", implying that this type of historiography is more interested in fields – or strata – of problems than in individual achievements (1959: 5).

Finally, the "new history" of the *Annales* school, a school of historiography that became dominant in French historical scholarship after World War II, also made use of archaeology as a methodological metaphor. In his classic *Apologie pour l'histoire ou le Métier d'historien* (1953), Marc Bloch proposed that historians cure themselves of their excessive preoccupation with narratives by attending to archaeologists' methods, and the model of postwar *Annales* scholarship, Fernand Braudel's *La Méditerranée à l'époque de Philippe II*, was of course based on the notion of different time strata (Braudel 1972–74).

Foucault himself seems to have been drawn towards the archaeological metaphor right from the start. Thus already in his early *Maladie mentale et personnalité* of 1954, he speaks of "neurosis [as] ... a spontaneous archaeology of the libido" (1954: 26), and in his *thèse complémentaire pour le doctorat des lettres*, "Introduction à L'Anthropologie de Kant", submitted in 1960, he writes of the "sedimentations" of Kant's work that call for an "archaeology of the text" (1960: 3–4). In fact, Foucault later even claimed to have adopted the notion of archaeology from Kant's "Fortschritte der Metaphysik". As a reading of this text shows, however, Foucault's claim that there one finds "the word, the text and the meaning to which I refer" (1971b), is not believable; neither the term nor any meaning even vaguely resembling Foucauldian archaeology appears in the *Fortschritte*.

In *Folie et déraison: Histoire de la folie à l'âge classique* (1954) (MC), Foucault speaks only *en passant* of the "archeology of that silence" to which madmen were reduced in and after the Classical age. It is only in the subsequent books, *Naissance de la clinique: une archéologie du régard médical* (1963) (BC), *Les mots et les choses: une archéologie des sciences humaines* (1966) (OT), and *L'archéologie du savoir* (1969) (AK), that Foucault begins to use 'archaeology' as the central label for his own approach. It thus seems appropriate to say that whereas the employment of the archaeological metaphor was by and large coincidental and arbitrary in Foucault's earliest works, it was only in the sixties that he came to regard it as sufficient to characterize his overall project. This assessment is confirmed by Foucault himself where he admits to having used this term "initially ... a bit blindly" or that "it seems to me that chance has not guided me badly" (1969: 23).

Interestingly enough, Foucault has never attempted to spell out the archaeological metaphor in any detail, confining himself rather to the 'translation' of 'archaeology' as "description of the archive" (*ibid.*). In order to give something of a first general overview over what Foucauldian archaeology of knowledge amounts to, it seems a useful exercise to point out some parallels between some of its central claims, on the one hand, and key aspects of archaeology (proper), especially more recent archaeological stratigraphy, on the other hand.

As one is informed by any textbook, archaeology typically is the study of prehistory, i.e. of periods from which no texts have remained. Archaeology is the study of the "archaeological record", i.e. of the "fossilized results" of human behavior. Furthermore, archaeology is neither the history of individual men, nor the history of individual fossils. First, archaeology "cannot aspire to be biographical, and archaeologists are excluded from the school of 'great man' history" (Childe 1962: 14). Second, rather than in the unique implement, the archaeologist is interested in "types" of objects that can be found in significant archaeological contexts. In other words, archaeology seeks to order arbitrary aggregates of objects into assemblages of fossils that belong temporally together. Assemblages of the same type encountered at different sites are called "cultures", and thus the archaeologist determines "culture sequences" by investigating successive levels of different stratified sites (*ibid.*, 19).

It is also characteristic of the archaeologist's work, that she must meticulously define and delimit series and classifications. This is because absolute chronology was – until the invention of radiocarbon dating – impossible, and is still only one of several forms of dating. In principle, the archaeologist "need keep only *archaeological time*. For archaeological time exhibits succession but not duration" (*ibid.*, 30). Classificatory principles include the function of objects, the ordering of objects made for similar purposes at different times according to their degree of efficiency and sophistication ("typological series"), as well as geographical, chorological distribution. The construction of typological series needs to take into account not only the possibility of differential rates of change, but also the correlations between several such series (Childe 1956: 70–74). This constructive procedure thus presupposes the distinction between the relative age of a given artefact in its typological sequence, its relative age as an element of a stratum, and its absolute age as its date of origin.

Foucauldian archaeology also deals primarily with prehistory, at least if we understand prehistory in the sense which is frequent in writings on the historiography of science where the prehistory of a science is often taken to be a discourse (or a bundle of discourses) that are eventually transformed into the science in question (e.g. Hacking 1975: 9; Kuhn 1970: 21). Or, put differently, the prehistory of a current science is a discourse that is not immediately recognizable as part and parcel of the modern discipline (Lepenies 1978: 443). Foucauldian archaeology is preoccupied foremost with prehistory in this sense, even though it relates somewhat critically to this very notion, and even though it does not confine itself to it. Furthermore, like archaeology proper, Foucauldian archaeology is no biographical history, and rejects the focus upon "great men". It is interested in defining series and types of objects, statements, discourses and concepts rather than concerned with individuals or unique achievements. It deplores the practice of many historians of science who take certain units like discipline, period, or oeuvre, for granted, and it calls for an explicit methodology for the construction of temporal series, discourses, their interrelations and correlations, as well as their rates of change. In so doing, it puts little emphasis upon dating scientific events by means of the calendar; rather, it seeks to define different time-scales in the history of science, different rhythms of change, and different relative ages of contemporaneous sciences. In distinguishing different levels, Foucauldian archaeology wants to show, for instance, that the social construction of objects like madness need not be contemporaneous with the genesis of a scientific discourse about these objects.

Even though these parallels between archaeology proper and Foucauldian archaeology already suffice to explain the naturalness of the metaphor in the latter's case, we can still go one step further by comparing Foucauldian archaeology with a more recent 'revolutionary' advance in archaeology proper, i.e. Edward Harris's "archaeological stratigraphy" (Harris 1979). This is captivating especially since Harris's criticism of traditional archaeology and its leaning towards geological stratigraphy has a counterpart in Foucault's criticism of the history of ideas, the philosophy of history and – at least indirectly – structuralism.

First, whereas Harris argues that archaeology needs a systematic theory of archaeological stratification, and whereas Harris accuses his colleagues of too often relying on a problematic conception of geological, i.e. natural, strata (1979: xiii), Foucault calls for a genuine *historical* theory of historical strata, and renounces the explicit or implicit employment of periodizations taken over from the philosophy of history. In like manner he distances himself from the structuralist-geological search for "the most majestic meaning of all" (Lévi-Strauss) by claiming that the archaeology of knowledge "does not imply the search for a beginning; it does not relate analysis to geological excavation" (AK 131).

The attack on the assimilation of the subject matter of archaeology to natural objects is also common to both Harris and Foucault in so far as both wish to emphasize the inanimate character of their basic 'fossils': Harris faults archaeologists for tacitly assimilating archaeological artifacts to geological, animate, fossils (1979: 5); Foucault in turn deplores the organistic metaphors that historians of ideas use when talking of the alleged "birth", "life" and "death" of ideas or themata.

Furthermore, whereas Harris sees the stratum as the basic unit to be reconstructed by archaeology, earlier archaeologists were often exclusively searching for the chief relics, artifacts or monuments (*ibid.*, 15). Similarly, Foucault is as outspoken a proponent as any of what has become standardly referred to as "vertical history" (e.g. Kragh 1987:

81). To refer to the vertical history of science as the history of strata is, of course, not unique to Foucault; e.g. David Knight writes that historians of science "have tended to drill a small hole down from the present through the strata of history; they should be well advised instead to look much more closely at the contents of one particular stratum" (Knight 1975: 25).

To drill "small holes down" is of course also a standard method in archaeology and geology where such holes are produced in order to obtain "sections" of given excavation sites. Harris is highly skeptical of this method, since it constitutes again a transferring of geological methods to archaeology. Since "most sites produce multilinear stratigraphic sequences of a complexity which would baffle many geologists ... it would be difficult to obtain a section which would be representative of little more than the section itself" (1979: 51). This shortcoming, Harris claims, cannot even be overcome by following the advice of one well-known textbook: "... look[..] at the section upside-down (standing that is, with the back to the section and bending to look through the legs); from this unaccustomed posture it is frequently possible to notice details not apparent to the normal view" (*ibid.*, 54). While I prefer to leave it to the reader to find a counterpart to this advice in the writings on the history or philosophy of science, it should be mentioned in any case that Foucault *mutatis mutandis* not only criticizes the historiographical practice of following just one idea or discipline through its history, but also denounces the tendency (for instance of Bachelard and Althusser) to draw conclusions about the development of all sciences on the basis of one or two, namely mathematics or physics. To put it differently, according to Foucault, the history of mathematics and physics does not provide us with a representative section for the histories of other sciences.

Of course, neither the archaeologist proper nor the archaeologist of knowledge can avoid the drawing of sections, even though, in both Harris's and Foucault's case, section drawing is only one method and needs to be supplemented by investigations into "boundary contours", i.e. vertical boundary lines of archaeological units. When sections are drawn, however, Harris calls for the drawing of "stylized sections" with "interfacial lines and its layers ... numbered" (1979: 58). (Realistic sections record only the artifacts found.) Likewise, *The Archaeology of*

Knowledge demands that the historian attend to discontinuities in the history of science, even if such discontinuities are never total and instantaneous.

This last parallel between Harrisian and Foucauldian stratigraphy can be strengthened by attending to three further points of contact. First, Harris calls upon archaeologists to take account of the fact that the usual order of geological stratification, according to which the strata lay upon one another in a natural order of age, does not hold true for archaeological stratification. Because of human intervention, or the intervention of natural processes, the order of strata need not correspond to age (1979: xiii). Second, and in part as a corollary, the archaeologist must pay special attention to the shapes of strata. She should try to excavate one stratum at a time rather than excavate arbitrarily or metrically, say, ten centimeters throughout the site (*ibid.*, 16). Third, archaeology needs a theory to explain the interfaces between strata, and this theory cannot be taken over unmodified from geology. Interfaces can just be boundaries of a given stratum ("period interfaces"), but they can also constitute "feature interfaces", i.e. strata in their own right (that perhaps have been squeezed into a thin layer by later depositions) (*ibid.*, 43).

Harris' attention to the order of strata and his stress upon the non-horizontal shape of strata interfaces coincides well with Foucault's emphasis upon the non-synchronic character of changes and transformations in or between discourses. For instance, two sciences can develop along similar patterns, yet this development can take much more time in the case of one as compared with the other. Furthermore, the different elements needed for the formation of a scientific discourse of some given type may well have different and distinct histories and they may be correlated and interrelated to form this discourse only in some specific span of time. The shape of the stratum to which this discourse belongs will thus possibly include these earlier elements. This implies that Foucauldian archaeology (at least in 1969) has to reject the arbitrary usage of period labels like "renaissance" or "Classical age", and that it refuses to have any of the metrical stratigraphy of George Sarton who once cut up the history of science in Europe and Asia in strata of fifty years (Sarton 1975).

The *Archaeology of Knowledge* also proposes an apparatus for the conceptualization of what appear to be breaks or discontinuities. Other than, say, Althusser or Kuhn, Foucault is not satisfied with the alternative 'total break versus continuity'. *The Archaeology* introduces a list of new terms and categories by means of which different forms of continuity and discontinuity can be distinguished.

Finally, and to conclude this comparison between Harris's stratigraphy and Foucault's archaeology, both reproach their respective colleagues for putting an exaggerated emphasis upon the dates of origin of their fossils (Harris 1979: 97). As important as is the origin of some artefact or statement, as important is, for both authors, its subsequent use, transformation and deposition.

3. THE NEW HISTORIES IN FRANCE

Few commentators have failed to notice the fact that *Annales* historiography and the historical epistemology of Bachelard and Canguilhem figure centrally in *The Archaeology of Knowledge*. Indeed, this fact is somewhat hard to overlook since the book mentions both schools repeatedly. What I regard as the main shortcoming of existing commentaries is rather that they focus on one of the two schools at the expense of the other, and that they do not pay sufficient attention to parallels in the concerns, theories and problems of the two. To set the record straight seems to be of some interest, not only as far as Foucault is concerned. Now that historians and philosophers of science have begun to study science from a more anthropological-cultural perspective, the work of *Annales* historians and of the epistemological school has to loom large; after all, in these fields of scholarship these allegedly new topics have been among the central preoccupations for the last fifty years.

THE *ANNALES* EXEMPLARS

In order to see the importance of the *Annales* School for Foucauldian archaeology in particular, and for the historiography of science in general, two central topics of this school of historians need to be attended to in more detail: the notion of mentality and the methodology of serial history.[4]

Against traditional history

The *Annales* School dates back to the turn of the century, roughly to the founding of the *Revue de synthèse historique* by the philosopher Henri Berr. Berr together with a group of like-minded historians and social scientists reacted to the challenge posed by Durkheimian sociologists who had denied history the status of a science and who had claimed the scientific study of society in its historical dimension to be solely the domain of sociology. In answering this challenge, Berr's group differed from more traditional historians. The more conservative historians belonging to *l'école méthodique*, like Charles-Victor Langlois and Charles Seignobos, and centered on *La Revue historique*, were ready to capitulate in the face of the sociological challenge, i.e. willing to concede that history as a discipline is nothing more than the study of the particular and that the scientific dimension of historical study consists exclusively in the application of a series of analytical methods of text-criticism (internal and external critique of documents). Unsatisfied with this deferential attitude, a band of historians grouping around Berr's new journal aspired to the status of a science by adopting the methods of those very sciences (sociology, geography, psychology, economics) that threatened history's traditional role.

It was in Berr's journal and in his monograph series *L'évolution de l'Humanité* that Lucien Febvre and Marc Bloch, the leading first-generation figures of what later became the *Annales* School, published their first studies, before founding, in 1929, the revue from which the name of the school derives, *Les Annales d'histoire économique et sociale*. Whereas before World War II the school was long denied access to the leading institutions in Paris (Febvre and Bloch worked in Strasbourg), after 1945 it became the leading school in French historical scholarship. In 1946 the journal was renamed as *Annales: Economies - Sociétés - Civilisations*, and in 1947 it acquired its own institutional bastion, the *Sixième Section* at the *École Pratique des Hautes Etudes*. The influence of the group soon became so pervasive that critics have come to speak of "the *Annalist holding*" (Bourdé & Martin 1983: 249).

What the *Annales* historians never tired of attacking was the preoccupation of the *école méthodique* and other forms of "traditional" history with producing narratives that concentrate on (mainly political) events and that are based, almost exclusively, on written documents. For Febvre, Bloch and their pupils, such narrative history neglects long-term or large-scale phenomena like material culture, trading routes, economic or demographic patterns of development, and mentalities. For instance, Fernand Braudel who inherited the leadership of the school from Febvre in the late fifties, renounces traditional history in the following way:

> Traditional history, with its concern for the short time span, for the individual and the event, has long accustomed us to the headlong, dramatic, breathless rush of its narrative. ... a short time span, proportionate to individuals, to daily life, to our illusions, to our hasty awareness - above all the time of the chronicle and the journalist (Braudel 1980: 27–28).

Marc Bloch, in his *Apologie pour l'histoire*, identifies another theme of traditional history: "the obsession with origins" (1953: 29). In searching for the ultimate origins of events and ideas, traditional history typically conflates beginnings and causes. It looks for the ultimate starting-points of events and ideas and neglects the historical strata in which these events and ideas are situated. Thus it "confuses ancestry with explanation": "As if the main problem were not to understand how and why the transition had taken place" (*ibid.*, 33). To provide an alternative, Bloch suggests that "a historical phenomenon can never be understood apart from its moment in time. ... As the old Arab proverb has it: 'Men resemble their times more than they do their father'" (*ibid.*, 35).

Bloch is also unconvinced by the key assumption of the *école méthodique*, according to which the historian obtains knowledge only indirectly, i.e. via reports of contemporaneous witnesses to historical events. It is here that Bloch proposes that historians would do better to take their lead from archaeology:

> If the best-known theorists of our methods had not shown such an astonishing and arrogant indifference toward the techniques of archaeology, if they had not been as obsessed with narrative in the category of documents as they were with incident in the category of actions, they would doubtless have been less ready to throw us back upon an eternally dependent method of observation (*ibid.*, 53).

Lucien Febvre criticizes not only traditional history in general, but also intellectual history, or the history of ideas in particular. He rejects the use of catch-all categories like feudalism or renaissance, and demands that historians reconstruct different systems of thought before advancing to generalizations (Chartier 1982: 16). Moreover, Febvre attacks, in harsh terms, historians of ideas who either follow singular, isolated ideas through the past, or attempt to analyze systems of thought in isolation from their larger cultural context:

> Of all the workers who cling to the generic title of historian, there is only one group that cannot in some way justify it in our eyes. They are those who, applying themselves to rethink for their own purposes systems that are sometimes several centuries old, and without the slightest care to show the links with other manifestations of the epoch which saw them come into being, end up doing precisely the opposite of what a historical method demands. They engender concepts from disincarnated minds which live their lives beyond time and space. Historians who deal with them forge strange chains whose links are at the same time unreal and closed. (Quoted from Chartier 1982: 17.)

The history of mentalities

However, Febvre distances himself from the Burckhardtian way of explaining similarities of contemporaneous historical phenomena in terms of a common *Zeitgeist*. Or rather, Febvre holds that the *Zeitgeist* itself stands in need of explanation (Chartier 1982: 18). In order to improve upon the traditional methodologies, Febvre as well as Bloch relied on a notion that had become central in the ethnology of Lucien Lévy-Bruhl, in the psychology of Henri Wallon (Le Goff 1985: 171), and that also appeared *en passant* in J. Huizinga's *The Waning of the Middle Ages* (Huizinga 1970). This notion is the notion of "mentality" (*mentalité*).

With what may be regarded as their respective major works, Bloch's *Feudal Society* (1936) and Febvre's *The Problem of Unbelief in the Sixteenth Century. The Religion of Rabelais* (1942), Bloch and Febvre provided the two classic *loci* of *histoire des mentalités* in *Annales* scholarship. Under the title "modes of feeling and thought" (Bloch 1962: 72–87), Bloch describes how medieval men and women experienced nature as threatening and unfamiliar, human life as endangered by disease, violence and the lack of hygiene, how they accepted the supernatural, why they could not doubt its existence, and how they related to myths, rites, theology, religion. Moreover, Bloch studies the passion with which emotions were expressed, how irrationalism surfaced both in the limited role left for reasoned arguments and in the importance given to dreams, how time was taken to be both circular and linear, and why human action was regarded as predetermined. Bloch also attends to the lack of precision in thought, a feature due in part to the absence of a developed system of numerals, the lack of honesty, the low degree of literacy, the strong role of dialects, and the ambiguity of words. Bloch concludes by commenting on the low esteem for learning and the typically low self-consciousness. In dealing with these elements of the collective mentality, Bloch remains close to anthropological methods, studies how these attitudes were expressed in everyday practices, and makes no concessions to individual differences (Burguière 1982: 433).

Febvre's *The Problem of Unbelief* runs a somewhat different course in that it takes its starting point from an individual, Rabelais, whose "limits of what is thinkable" Febvre seeks to determine by placing him within his time. Even though Febvre is keenly interested in feelings, sensibility and emotions, his approach to the mentality of sixteenth century France is by and large slightly more intellectualist than Bloch's. This difference of emphasis is clear from the fact that for Febvre the core of a mentality is its "mental tools". These tools are first and foremost "the ensemble of categories of perception, conceptualization, of expression and action which structure individual as well as collective experience" (Revel 1986: 451). With respect to the mentality of the sixteenth century, for example, Febvre argues that the *outillage mentale* was marked by a lack of crucial words, like "absolute", "relative", "abstract", "concrete", "complex", "confused", "adequate", "virtual", "intentional" or "transcendental"; by the absence of a nomenclature; by a syntax which made the expression of perspective rather difficult and which did not clearly distinguish between co-ordination and subordination; by a low degree of formalization in the sciences and mathematics; and by the inaccessibility of important books. Febvre explains the prevailing attitudes and beliefs of the sixteenth century in terms of these mental tools. It is in this manner that he investigates religion (assumptions about God, soul and immortality, the possibility of doubt and choice in religious matters), logicality (the treatment of conflicts in intellectual questions, the awareness of contradictions, modes of reasoning, requirements of proofs), time (types of chronometers, the historical sense), images of sciences and the scientist (compilation vs. explanation, dogmatism vs. skepticism, veracity, professional images, the relation between theoretical knowledge and practical skill, the role of observation), cosmology (ways of reasoning in questions of cosmology), emotions (ways of expressing one's feelings), senses (the role of different senses), art (which genres were central), and the category of the possible (to what extent are physical and logical impossibility held distinct).

In some places, Febvre seems to suggest that two historical periods with diverging mental tools are incommensurable: "... there is really no common measure between sixteenth-century man's way of feeling, thinking and speaking and our own" (Febvre 1982: 100). This viewpoint motivates Febvre's accusation of massive anachronism directed against those historians of ideas who have tried to present Rabelais as an atheist, and who have made no distinction between the atheism of the sixteenth and later centuries. Against these earlier interpretations, Febvre seeks to show that there is, as one recent commentator puts it, "an epistemological break, a conceptual caesura between the unbelief of the sixteenth and that of the eighteenth century" (Wootton 1988: 703). In other words, Febvre replaced "an idea of incremental progress with one of radical discontinuity" (*ibid.*, 729). Thus in his *Rabelais*, Febvre tries to show that for the sixteenth century thinker atheism meant merely deviation from some specific doctrinal elements of christian religion rather than disbelief in the supernatural.

While strongly advocating radical breaks in the history of thought, Febvre does not hold, however, that mental tools change overnight:

> Every civilization has its own mental tools. Even more, every era of the same civilization, every advance in technology or science that gives it its character, has a revised set of tools, a little more refined for certain purposes, a little less so for others. A civilization of an

> era has no assurance that it will be able to transmit these mental tools in their entirety to succeeding civilisations and eras. The tools may undergo significant deterioration; or, on the contrary, more improvement, enrichment, and complexity. They are valuable for the civilization that succeeds in forging them, and they are valuable for the era that uses them; they are not valuable for all eternity, or for all humanity, not even for the whole course of development within one civilization (Febvre 1982: 150).

Even though Febvre thus allows not only for the development of mental tools but also for their being (at least in part) inherited by subsequent eras, ever since Febvre's and Bloch's classics, the *histoire des mentalitès* has continuously been haunted by the problem of accounting for change and development. Emphasizing strongly the pervasive influence of prevailing mentalities, the *histoire des mentalitès* has been unable to distinguish between the change within one given mentality and the replacement of one mentality by another (see e.g. Burke 1986: 443; Chartier 1982: 31; Wootton 1988: 703). Since they tend to conceive of mentalities as "prisons" (Braudel 1982: 31), *Annales* historians make it difficult to understand how any culture could ever move on to 'inhabit' a new one.

Nevertheless, some suggestions concerning the problem of change have been made. For instance, as the above quotation shows, Febvre leaves room for the possibility that mental tools change piecemeal. Some passages of *The Problem of Unbelief* also suggest that Febvre reckons with variation *within* a given mentality; some groups, like scientists or artists, make full use of the available "idea materials", while others are able to employ only a small part of the mental tools of their time (Chartier 1982: 21). Furthermore, and interestingly enough, in order to cope with change, historians of mentalities permit the history of ideas reentrance into the *Annales* paradigms. For example, Jacques Le Goff, who together with Emmanuel Le Roy Ladurie inherited the leadership of the school from Fernand Braudel in the early seventies, writes that "the mentality of any one individual ... is precisely what that individual shares with other men of his time" (Le Goff 1985: 167). This characterization leaves room for individual originality and difference as an area of thought not shared with others. Whereas Le Goff proposes historical psychology ("archaeopsychology") and ethnology as the proper approaches for the study of mentalities, he provides space for the history of ideas as the study of great thinkers. The study of mentalities and the history of ideas have to be related, however, for instance in order to investigate the origins of elements of a given mentality:

> The history of mentalities must also be distinguished from the history of ideas, out of opposition to which it also grew. It was not the ideas of Thomas Aquinas or Saint Bonaventura which, from the thirteenth century on, governed people's minds, but mental *nebulae* in which such ideas played only a part as deformed echoes, devalued fragments and words taken out of context (1985: 176).

Finally, in a book that otherwise closely follows the outline of Febvre's *opus magnum*, R. Mandrou's *Introduction to modern France 1500–1640. An essay in historical psychology* (Mandrou 1975), mentalities are made more dynamic by introducing different time scales. Thus Mandrou distinguishes between long-term "common features ... as the basic features common to every mentality at the time" (*ibid.*, 236), i.e. features like hypersensitivity of temperament, social aggressiveness, and man's feeling of impotence in the face of the natural world; "views of the world: mental structures" as the ensemble of "all the notions – intellectual as well as ethical – within which individuals and groups daily developed their thoughts and actions" (*ibid.*, 243) (at this level there are numerous differences between various groups, classes and individuals); and "climates of sensitivity: mental conjunctures": these are periods of shorter or longer duration, from a few years up to a generation, periods of optimism or pessimism, of crisis or confidence. According to Mandrou, it is at this level that "gradual changes in outlook" originate (*ibid.*, 249).

Serial history and the construction of the historical fact

Mandrou's distinction between different time scales within a mentality is of course but an application of quantitative or serial history to the *histoire des mentalités*. The differentiation between different time-scales does not come from Febvre or Bloch but from the work of F. Simiand and C.-E. Labrousse, two economic historians who exerted a strong influence on the *Annales* School. Nevertheless, the *locus classicus* within *Annalism* for the distinction between different times is the monumental *La Méditerranée* of Braudel. Since we have already encountered his characterization of the time of events as the time of the short time span, it remains to mention his two other time strata, the time of cyclical movements or conjunctures, and the time of structures, i.e. the *durée longue*. The time of cyclical movements is the time of "a price curve, a demographic progression, the movement of wages, the variations in interest rates, the study (...) of productivity, ... [or of] money supply" (Braudel 1982: 29). Structures *qua* phenomena that persist over centuries function as "hindrances ... [or] limits (...) beyond which man and his experiences cannot go. Just think of the difficulties of breaking out of certain geographical frameworks, certain biological realities, certain limits of productivity, even particular spiritual constraints ..." (*ibid.*, 31). Braudel divides *La Méditerranée* in accordance with this three-partite distinction. However, even though he argues that "these different time spans ... are all interdependent" and that "... to be able to achieve an imaginative understanding of one of these time spans is to be able to understand them all" (*ibid.*, 48), he in fact does not succeed in explaining the interrelations between these time spans.

While the works on structures and conjunctures within *Annales* scholarship make for fascinating reading, I shall not try to summarize the numerous studies that the *Sixième Section* has carried out in the field of the quantitative history of conjunctures of different durations, studies, for instance on prices, salaries, trade, demography, epidemics or the climate. For our purposes it is more important to take note of the new historiographical philosophy that takes shape in these studies. This new view of the nature of historical research comes out especially clearly in a remark by Emmanuel Le Roy Ladurie in his *Times of Feast, Times of Famine: A History of the Climate since the Year 1000* (Le Roy Ladurie 1972) and a programmatic article by François Furet (Furet 1985).

Le Roy Ladurie proposes that the new serial history is no longer bound to "the anthropocentric approach" (1972: 16). Serial history can be applied to any series of data, regardless of whether these data involve – directly or indirectly – human beings. As a consequence, Le Roy

Ladurie no longer underwrites Marc Bloch's famous claim, according to which "behind the tangible features of landscape ... it is human beings that the historian is trying to discern" (*ibid.*, 18). As the historian of the climate sees it, it "is mutilating the historian to make him into no more than a specialist of humanity ... he may also in certain cases be interested in nature for its own sake, and make known by his own irreplaceable methods nature's own special Time" (*ibid.*, 20). What makes this thesis remarkable is that with it *Annalist* criticism of the traditional political event reaches its peak. Whereas the first and the second generation of the school 'deconstructed' the event by enlarging the time scale, Le Roy Ladurie goes one step further and challenges the connection between history and the human being. Needless to say, such a viewpoint was well received at a time when antihumanism loomed large, and Foucault announced the "end of man".

The programmatic article of François Furet that I referred to above is his "L'histoire quantitative et la construction du fait historique". This article was published in the *Annales* in 1971, and later translated into English as part of a collection of articles by *Annales* historians, fittingly entitled *Constructing the Past*.[5] And indeed, it is this aspect of "construction" that succinctly sums up the new historiographic attitude.

Furet's starting point is the distinction between quantitative and serial history, a distinction first used by Pierre Chaunu (e.g. Chaunu 1970). Whereas quantitative history refers to any kind of history that employs mathematical tools, serial history (*histoire sérielle*) is interested in the quantification of repeated facts; it constructs "temporal series of homogeneous and comparable units" in order to measure "the evolution of that reality at given intervals, usually annually". In this way, serial history, right from the start, rejects the traditional historian's preoccupation with the individual fact. Its subject matter is not found but constructed; it is not the unique, but the repeatable. Historical reality, or the historical record, can be broken up or constructed in several ways and at different levels, and it is up to the historian to establish correlations between these different levels (Furet 1985: 14).

The serial historian must also distinguish between change and mutation, must define thresholds and breaks: as long as cyclical patterns of development prevail, the serial historian is led to speak of "change within stability". Yet when the serial historian encounters, for instance, tendencies towards indefinite growth, she must assume that the latter may alter "the very structure of time and rhythms of change". In order to be able to distinguish the later sort of mutations from changes within stability, i.e. in order not to assimilate mutations to cyclical changes, the serial historian must work with temporal units of various length (*ibid.*, 15).

Another problem that the serial historian must cope with is the fact that European archives have been set up mainly for the study of political events rather than for the study of long-term developments ("an archive is the memory of nations"). The big European archives contain the documents that the traditional historian treats as evidence for, or testimony of, historical facts: "In this sense, a historical document is a unique, discrete, particular moment within a global evolution which either remains temporally indeterminate or is divided into centuries, reigns, ministries." For the serial historian, on the other hand, the elements of the historical record are no longer documents about the 'significant turns in history'. External criticism, like the comparison of a document with other contemporary texts, is no longer the most crucial historiographic operation. Rather, what becomes central is the comparability and homogeneity of a given text with other texts in the same temporal series (*ibid.*, 16).

Furthermore, the document or text is no longer treated as a unit of meaning in and by itself. The serial historian rejects the idea according to which "the sources ... define the questions asked by a discipline"; instead, it is the historian's query that "determine[s] the sources" (*ibid.*, 18). For instance, rather than asking for the meaning of texts, the serial historian can investigate handwriting and thus arrive at statistically obtained conclusions about the development of literacy (*ibid.*, 17).

These new procedures and methodological problems of serial history, Furet holds, amount to "a revolution in historiographical consciousness" (*ibid.*, 21). The core of this revolution is the idea of construction. The essence of the matter is not that serial historians *construct* their subject matter, whereas more traditional, document or event-oriented history *found* it. Rather, the emergence of serial history, and the discussion of its methodology, has brought to the fore that *all* historical research is inevitably constructive:

> ... today's historian must give up methodological naivety and think about the conditions under which historical knowledge is established. ... the mask of some kind of historical objectivity hidden in the 'facts' and discovered at the same time as them, has been removed for ever; the historian can no longer avoid being aware that he has constructed his 'facts' ... (*ibid.*, 20).

More specifically, Furet claims, the coming of age of serial history implies a transformation of the conception of time in history. For traditional history with its prejudice in favor of the unique and irreplaceable, an event cannot be placed in a series of comparable, measurable data. It follows that an event is only studied in the short term perspective or that it is interpreted in terms of a teleological meaning: "it has no past but it can have a future" (*ibid.*, 21). The teleological meaning in question has been the progress towards modern forms of the state or modern forms of rationality and reason. Moreover, even when events were related to one another in traditional history, this relating typically happened in terms of a unifying meaning, like *vision du monde* or *Zeitgeist*, that was postulated independently of the events. Looked at from the angle of temporality, these commitments of traditional history imply a curious way of understanding historical time: a series of unrepeatable and unique events is narrated in the form of a story whose plot has been pre-established on philosophical and ideological grounds; in Furet's terms: "a series of discontinuities described in the continuous mode - a story" (*ibid.*, 21).

The conception of historical time involved in serial history, however, is diametrically opposed to that of traditional history: "Serial history ... describes continuities in the discontinuous mode: it is history as problem rather than history as story" (*ibid.*, 21). Even though serial history starts from constructing homogeneous series (*qua* continuities), it distinguishes between different levels, thresholds and cycles (*qua* discontinuities). It works with series that have different speeds and temporal rhythms, it identifies series not all of which can be correlated. Serial history also compares cycles taking place at different times or in different regions: in the first case, similarities may point to common geographical factors, in the latter, nonsimilarity may suggest difference in economic structures (*ibid.*, 24).

Taking the advances of serial history seriously demands then, according to Furet, that history "must now start from the hypothesis that the nature of historical time may be different according to the level of reality or the partial historical system which is being analyzed" (*ibid.*, 25). From this it follows, first, that the tacit assumption of traditional history, according to which different elements of a historical period develop approximately in parallel, has to be kept in suspense. Rather than start from a given periodization, like the Renaissance, and study different elements within this period, the historian is better advised to investigate elements over a much longer period. And second, only when using such a procedure, Furet claims, can the historian answer the question as to which elements and patterns of a given period undergo rapid development and which are marked by stagnation (*ibid.*, 26).

The Exemplars of Historical Epistemology

Duhem vs. Koyré

The study of the history of science in France has been decisively shaped by the famous disagreement between Pierre Duhem and Alexandre Koyré over whether the advances in the natural sciences during the sixteenth and seventeenth centuries qualify for the title of a scientific revolution. Duhem repeatedly argues that the answer should be negative by pointing out that central concepts of Galilei can already be found in the writings of the Parisian masters of the fourteenth century:

> The mechanical and physical science, of which modern times is rightfully so proud, derives, in a hardly perceptible manner, from an uninterrupted series of improvements of doctrines professed in the womb of medieval schools; the alleged intellectual revolutions were most often nothing but slow and long-prepared evolutions; the self-proclaimed renaissances frequently nothing but unjust and sterile reactions. Respect for tradition is an essential condition of scientific progress (Duhem 1905: iv).

The development of mechanics must be declared "an evolution; each of the stages of this evolution is a natural corollary of the stages which precede it, and a part of the stages that follow it" (quoted from Fichant 1969: 73). This observation leads Duhem to oppose talk of scientific revolutions as "brisk jumps" (Duhem 1906b: 156). While there can be acceleration of the speed of development, this occasional acceleration is simply a "condensation" of the evolution (Fichant 1969: 74). Indeed, according to Duhem, the study of the history of science is usually distorted by "two prejudices", to wit, that scientific progress is but "a sequence of sudden and unforeseen discoveries", and that it is "the work of geniuses who have no precursors at all" (Duhem 1906a: 1). For Duhem, a new physical theory is "not the sudden product of a creation, it is the slow and progressive result of an evolution" (1906b: 365). Moreover, scientific theory does not originate from a sudden break with prescientific thought: "in the genesis of a scientific doctrine there is no absolute beginning"; it will always be possible to identify earlier opinions which have "prepared, suggested and announced" the theory in question (1913: 5). Duhem also believes in the existence of eternal themes and ideas, for instance, the questions concerning the value of physical theory and its relation to metaphysics "belong to all times" (1969: 3).

Duhem's downgrading of Galileo's achievements is informed by his strong catholicism and his positivism. Or rather, his positivism – the sharp distinction between physical-mathematical theory and metaphysics – is his device for "saving" the achievements of medieval scholasticism. This religious prejudice comes to the fore not only in peculiar asides like "Kepler was a Protestant, but [!] deeply religious" (1969: 100), but also in Duhem's overall strategy of replacing the scientific by a "theological revolution". As he alleges, it was the Church Fathers' attack on the pagan philosophers, their rejection of the notion of eternal prime matter and the belief in the domination of planets over sublunary events, that cleared the ground for modern science; "it was necessary that a theological revolution take place", and this revolution was "the work of Christian theology" (1914: 453). Thus, what Duhem denies the scientists, he readily concedes to the theologians: the making of a revolution in thought.

Koyré, on the other hand, holds that "the apparent continuity in the development of physics, from the Middle Ages to modern times ... is illusionary" (1973: 172). He speaks of the advances of the sixteenth and seventeenth century as "a deep revolution which changed the framework and patterns of our thinking" (1957: vii), and allows in general for "veritable 'mutation[s]'" in science (1978: 1). Koyré studied the history of science in its interrelations with philosophical thought. For him the scientific revolution meant the dissolution of the Greek and medieval conception of the cosmos:

> ... the founders of modern science ... had to destroy a world and replace it with another. They had to reform the structure of our very intelligence, reformulate and revise its concepts, envisage Being in a new manner, elaborate a new concept of knowledge, a new concept of science – and they also had to substitute a quite natural point of view, i.e. that of common sense, with another [point of view] that is not at all natural (1973: 172).

Despite this direct disagreement with Duhem, and despite his criticism of the pervasive use of the categories of precursor and influence in Duhem (Koyré 1957: 5–6), Koyré does not depict scientific revolutions as abrupt ruptures. For example, Koyré is ready to write that Kepler anticipated modern empirism and positivism, and he defends this claim by arguing for continuity with respect to basic positions in epistemology (*ibid.*, 58, 286). He also presents Archimedes as "the veritable precursor of modern physics" (1973: 172), and interprets the Newtonian synthesis of mathematical physics as the result of a "resurrection" of Democritos' atomism and Plato's mathematics (*ibid.*, 333). For Koyré, revolutions too need preparation, they "too, have a history". In fact, the destruction of the cosmos and the geometrization of space, the two central elements of the scientific revolution, needed about 150 years to be accomplished (1957: ix). Finally, Koyré also allows for the non-uniformity of progress and the non-homogeneity of contemporaneous works:

> Palingenius and Copernicus are practically contemporaries. Indeed, the *Zodiacus vitae* and the *De revolutionibus orbium coelestium* must have been written about the same time. Yet they have nothing, or

nearly nothing, in common. They are as far away from each other as if they were separated by centuries (*ibid.*, 28).

Bachelard

Gaston Bachelard's epistemology emphasizes revolutions, breaks or discontinuities in the history of the natural sciences even more than Koyré. However, Bachelard is less concerned with the sixteenth and seventeenth centuries than with the physical "revolutions" in our century. It was the period in physics from Einstein to Chadwick, i.e. the period of the special and general theory of relativity, wave mechanics, quantum mechanics, the discovery of cosmic rays, the probabilistic theory of the electron, the uncertainty principle, the relativist interpretation of wave mechanics, the discovery of neutrinos, positrons, neutrons and mesons, which convinced Bachelard that "one decade in our epoch is equal to centuries in earlier epochs" (Lecourt 1975: 32, 95), and that this new physical science is "without antecedents" (*ibid.*, 36).

Bachelard is especially fascinated by the fact that microphysics breaks radically with common sense. Whereas in earlier times the natural scientist was "one of us", and whereas in earlier centuries "natural science and philosophy spoke the same language" (Bachelard 1971: 16), modern physics forces us to adopt a completely new ontology in which objects are no longer basic. This new ontology also implies "an absolute break" with the conceptions and the "reism" of classical physics (*ibid.*, 59).

Modern physics is characterized by Bachelard with notions like "noumenology", "active empirism", "phenomenotechnique", "rational materialism", and "applied" or "dialectical rationalism". The common denominator of these various labels is the idea that twentieth century physics does not start from observation but from "an organization of *objects of thought*" that only subsequently "become *objects of technical-experimental experience*, in a pure artificiality of experience" (*ibid.*, 62). In other words, the objects of physical science are constructed by experimental means, not found; modern physics is "a science of *effects* rather than facts", "experience is no longer a starting point, it is not a simple guide any more, experience is a *goal*" (*ibid.*, 75).

The experimentalist is not, however, the mere handyman of the theoretician. Rather modern physicists have an "*abstract-concrete* mentality" (*ibid.*, 122) that allows challenges to go both ways: the experimentalist will search for the vindication of a phenomenon which the theorist announces as a possibility, whereas the theorist will, in the light of new experimental phenomena, modify her theory in order to show that her theory should have predicted the phenomenon in question (*ibid.*, 123). Bachelard justifies the assumption of a "dialectics" between theory and experiment by pointing out that the instruments of modern physical experimentation are but "reified theorem[s]" (*ibid.*, 149).

What gives Bachelard's reflections upon the history of modern physics in particular, and the development of science in general, their specific "epistemological" dimension, is first of all his thesis that different forms of scientific practice involve different epistemological views, and that the order in which the discourse of physics has gone through these positions provides us with "a principle of classification for different philosophies and for the study of the progress of reason" (*ibid.*, 34).

In *La philosophie du Non* (Bachelard 1978), Bachelard explains this classification in terms of the notion of mass. For the child mass is something which is judged in terms of size, "mass is only a *quantity* if it is fairly big" (*ibid.*, 36). This is the level of "animism". A higher level, that of "realism", is reached when the notion of mass is linked to the use of a scale. In this stage, thought is "empirical, solid, clear, positive, [yet] immobile" (*ibid.*, 39). The Newtonian notion of mass represents rationalism, since here the absolute use of the notion is replaced by a correlative use of several concepts. Mass is now defined in terms of other concepts and no longer regarded as an immediate element of direct experience. This marks the transition from "a realism of things to a realism of laws" (*ibid.*, 41). With the theory of relativity, we attain the level of "complex realism"; now the earlier belief that the mass of an object can be determined independent of its speed has to be given up; the notion of absolute mass turns out to be meaningless. Furthermore, mass is no longer independent of energy (*ibid.*, 44). Finally, with Dirac's mechanics we arrive at the stage of "dialectical surrealism", the stage where we are dealing with two masses for one object: "One of these masses comprises in a perfect way everything which the four earlier philosophies knew about mass ... But the second mass, the dialectical counterpart of the first, is a negative mass" (*ibid.*, 48).

In *Le rationalisme appliqué* (1949), Bachelard's "philosophical spectrum" differs somewhat; here he presents (naive) realism, empirism and positivism, on the one hand, and idealism, conventionalism and formalism, on the other hand, as two diametrically opposed series of philosophies, both of which are unable to grasp the nature of modern physics as "applied rationalism": formalism, conventionalism and idealism represent a series which increasingly loses the empirical element of science, while positivism, empirism and (naive) realism are successive stages of losing the constructive-experimental and mathematical-theoretical side (Bachelard 1971: 127).

While castigating these various epistemologies as inadequate philosophical stands, Bachelard holds that in everyday life as well as in science, we all are philosophical pluralists. For instance, in different situations we employ different notions of mass, i.e. we rely tacitly on different philosophies. Bachelard suggests that we can draw "epistemological profiles" of our understanding of scientific terms, profiles which depict the extent to which we abide by different philosophies (1978: 57).

In order to clarify transitions from one epistemological stage to the next, Bachelard introduces the dichotomy "epistemological obstacle" *versus* "epistemological act". Epistemological obstacles are what hinders the scientist in advancing to the next stair of Bachelard's epistemological staircase, while epistemological acts are successful steps made on the way up the staircase. Bachelard never provides any detailed typology of different epistemological obstacles, even though in *La formation de l'esprit scientifique* he distinguishes between "first experience", the "realistic obstacle", the "animistic obstacle" and "the libido". The first experience constitutes an obstacle in so far as its vividness and manifold character might hinder us to pose scientific questions; realism fixes our attention on things and substances; animism tempts us to treat active substances as animate; and the libido is responsible for our inclination to depict natural processes in terms of sexual metaphors (Bachelard 1971: 178–86).

By attending to different uses of the notion of "epistemological obstacle" in Bachelard's many books, one can distinguish at least the following cases: (1) "the primary experience" (the object of knowledge is taken to be immediately given in itself); (2) "the substantialist obstacle" (the assumption that the order of the real mirrors the order of thought; thus, e.g. the subject/predicate scheme is imposed upon nature as the substance/accidens scheme), (3) "general knowledge" (a system of concepts is composed based on vague analogies); (4) "pragmatic knowledge" (non-scientific interests and inclinations); (5) "the animist obstacle" (confusion of regions of objects, e.g. between physics and biology); and (6) "the quantitative obstacle" (the overestimation of quantification and measurement) (Brühmann 1980: 143–44).

These various forms of epistemological obstacles also provide *via negationis* some content for the opposite concept, the epistemological act, of which Bachelard only tells us that they are "those blows of scientific genius that bring unexpected impulses in the course of scientific development" (1971: 213). By means of epistemological acts the scientist cuts herself free from various inclinations and prejudices. In this sense, the epistemological act constitutes a leap forward to a level of thought that comes closer, or ultimately reaches, the stage of applied rationalism. More precisely, in escaping the epistemological obstacles, the scientist develops a system of concepts and experimental techniques, in other words, a "problematique", which by an internal dynamic – which Bachelard leaves unanalysed – leads to ever new questions, radical reorganizations of theories, and new instruments (Brühmann 1980: 144; Bachelard 1971: 32, 131, 135).

The dichotomy "epistemological obstacle *vs.* epistemological act" also shapes Bachelard's views on the issue of continuity *vs.* discontinuity in history, and thereby his conception of how the history of science should be written.

To begin with, Bachelard's above-mentioned stress on discontinuity is hardly surprising given his belief that decisive scientific advances involve epistemological acts, on the one hand, and a re-organization of systems of concepts and experimental techniques, on the other hand. However, for the epistemologist these discontinuities do not make talk of progress impossible. Of course, using the philosophical spectrum as an indicator of progress would be circular, since this spectrum has after all been derived from the factual development of physics. What allows Bachelard to stick to the perspective of progress despite the existence of discontinuities is rather his belief that the rejection of a past science does not rule out the preservation of its results in the form of special, limited cases of later theories:

> Non-Euclidean geometry was not invented in order to contradict Euclidean geometry. ... First constructed in the margins of Euclidean geometry, non-Euclidean geometry sheds a revealing light on the limitations of its predecessor. The same may be said of all the new varieties of scientific thought, ... (Bachelard 1984: 8).

Put differently, Bachelard denies that the new knowledge can be expressed in the old, rejected framework ("... one must not hope to find a simple formula for converting the new doctrines into terms comprehensible within the framework of the old", (*ibid.*, 8)), but he allows for the old knowledge to be expressed in the new. Perhaps it is this asymmetry which also informs his opposition to the categories of influence and the predecessor. Bachelard rejects as fruitless - at least for modern science - to speak of the influence of major ideas and discoveries across the borders of problematics (1971: 203), and he castigates attempts to make Hegel the precursor of Maxwell, and Raspail the forerunner of Bohr (see Lecourt 1975: 84–85).

Rather than search for influences and anticipations, the epistemological study of the history of science must construct two series of events, the series of epistemological acts and the series of epistemological obstacles. The former is the constructed object of "recurrent" or "accepted history" (*histoire sanctionée*), the latter the constructed subject matter of "obsolete history" (*histoire périmée*) (Bachelard (1951: 24–27). Neither of these two forms of writing history confines itself to a neutral narration of the history of science dated by the calendar; rather, both amount to an evaluative reconstruction of that history. Accepted history reconstructs history in terms of how it should have happened, obsolete history in terms of how it should not have developed. Accepted history identifies the steps that approach the truths of present-day science, obsolete history studies those elements that have no place in the teleological scheme of accepted history.

Finally, Bachelard's profile of philosophical attitudes involved in science enables him to speak of different rhythms and times of sciences, depending upon the rates at which these sciences cross the various thresholds. By the same token, the historical epistemologist can also propose that two sciences that are contemporaries when judged by the calendar, are not contemporaneous when judged by their epistemological profile (Brühmann 1980: 157): "Science is like a half-renovated city, wherein the new (the non-Euclidean, say) stands side by side with the old (the Euclidean)" (Bachelard 1984: 7).

Canguilhem

Bachelard's student and successor as the director of the Institut d'Histoire des Sciences et des Techniques at the University of Paris, Georges Canguilhem, continues on several of Bachelard's central lines while taking exception to some others. To begin with those aspects of the epistemological program which Canguilhem underwrites or pushes further, Canguilhem shares his teacher's conviction that "the history of science is the history of the defeats of irrationalism" (Canguilhem 1968: 183). He stresses the importance of the epistemological obstacle notion as a means of avoiding the illusion that progress is linear and that the history of science can confine itself to the recording of results (*ibid.*, 177). Canguilhem too holds that different sciences have different rhythms of development and different speeds at different times, and that the "epistemology of continuity" is insufficient and needs to be complemented by an "epistemology of discontinuity" (1988: 16). For instance, Canguilhem suggests that biophysics and biochemistry constitute a radical break with earlier biology. A crystal of DNA is "a 'superreal', nonnatural object, the product of considerable technical and theoretical labor ... of microextraction and microdissection, of combinatorial algebra, of statistics, of electron microscopy, of enzyme chemistry". In order to arrive at this new object, central elements of what formerly were taken to be crucial features of living things, like sexual reproduction and enzymatic reactions, had to be left aside. The new object of research was inconceivable for earlier biology, and its constitution in and through these new various techniques amounts to the "conversion to a new view of the world" (*ibid.*, 117–18).

In agreement with Bachelard, Canguilhem also emphasizes the

importance of problematics. Thus he argues, for example, that Mendel's discovery of the laws of heredity could not have hastened progress in biological research even if they had been recognized by scientists of the day. At the time, Mendel's work could not have been linked with Darwin's theory of variations: interest in evolution was then running high, and Mendel's laws would have appeared to his contemporaries as an outdated form of fixism (*ibid.*, 116). Furthermore, Darwin's work could not have been combined with a chemical study of the cell nucleus in a single problematic, since unicellular species were not considered as organisms, and since no-one could yet conceive of the possibility that the laws of life might be discovered by studying bacteria and microbes, i.e. discovered in the very threats to life itself (*ibid.*, 115).

Yet another aspect of Bachelard's epistemology that Canguilhem develops at some length is the criticism of the search for predecessors. For Canguilhem, the "virus of the predecessor" destroys the history of science, since "if there really were precursors, then the historical dimension of science would be an illusion" (1968: 21). It is the lack of epistemological critique, of analyses of problematics and obstacles which result in the anachronistic construction of a few central lines of influence. Such a relation, i.e. the relation of influence, between two authors may be assumed only once epistemological critique has shown that these authors have identical questions as well as research strategies and that the leading concepts of their respective systems of concepts are identical (*ibid.*, 22).

Other than Bachelard, however, Canguilhem does not employ the normative scheme of the "philosophical spectrum" and "epistemological acts" of genius, and he focuses more on social factors, ideology and the interrelations between disciplines. Canguilhem will have none of Bachelardian "strokes of genius" not only because the work of every scientist is always a mixture of the old and the new (1988: 15), but also because the individual scientist is usually dispensable. Had Darwin died during the voyage of the *Beagle*, his discoveries would have been made by Alfred Russel Wallace (*ibid.*, 110–11).

An extensive use of the notion of the flash of insight easily leads to a conceptualization of the history of science as a series of accidents. This is a viewpoint, however, which Canguilhem seeks to exorcise from history. He argues, for instance, that the discovery of iodine by Clémont and Desormes in 1812 was not - as is often thought - the result of an accident. Chemists at the time were searching for active substances present in organic compounds; thus even if a Parisian manufacturer of saltpetre had not brought the unknown substance which caused extensive corrosion to Clémont and Desormes, iodine would have been discovered sooner or later anyway (*ibid.*, 274–94; see also Lecourt 1975: 169–70).

Canguilhem's attention to the interplay among disciplines and the interrelations between sciences and society comes out clearly in his remarks on the history of statistics as well as in his two best-known studies, *Le Normal et le pathologique* (1943) and *La formation du concept de réflexe aux XVIIe et XVIIIe siècles* (1955).

The emergence of statistics, we are told, cannot be derived from the mathematics of Laplace, the biology of Darwin, the psychophysics of Fechner or the ethnology of Tylor. Biometrics and psychometrics become possible once new objects of observation became accessible outside of science. Quetelet's statistical study of human extremities became possible through the creation of national, standing armies, and Binet's study of intelligence was made possible by the introduction of elementary schools. Put differently, "the history of science is not dealing with merely a group of sciences without inner relations among them, but it is also dealing with non-science, with ideology and political and social practice" (Canguilhem 1968: 18–19).

In his study of different notions of disease in nineteenth century medicine and physiology, Canguilhem demonstrates that the shift from a quantitative notion of disease (disease as merely gradually different from the normal, healthy state) to a qualitative notion of disease (disease as principally different from the healthy state) must take into account not only medicine but also sociology. This is because the quantitative notion of disease and normality is first taken over by Comtean sociology, transformed for sociological purposes, only to re-enter medicine later in the century, where it is then soon replaced by the qualitative conception (Lepenies 1978: 446).

Finally, in his investigations into the origins of the notion of reflex, Canguilhem provides his reader with an excellent example of his theses that epistemological critique must carefully construct its historical trajectories, pay special attention to concepts, and not be misled by superficial similarities (cf. Lecourt 1975: 171–77; Gutting 1989: 35–37).

Canguilhem seeks to reject the assumption that a *description* of reflex movements within a mechanistic biology suffices to attribute the *concept* of reflex to Descartes. This is because, for Canguilhem, a concept is more than a description. It provides an interpretation of data that makes fruitful scientific work possible. In the case of reflex motion, the concept must include the ideas that the two movements from the sense-organs to the response-center and back are alike, and that the brain is not directly involved. Neither of these two decisive elements are part of Descartes's account, however. As Canguilhem shows, the concept of reflex motion originates instead within a framework which is foreign to mechanical biology. The concept of reflex has its origin in the work of Thomas Willis and Georg Prochaska, two vitalists of the seventeenth and eighteenth century. Since they likened life to light, they could model reflex motion on the reflection of light and conceive of a symmetry of the two movements involved. Furthermore, since they distinguished between cerebellum as the center of involuntary movement and cerebrum as the brain, they could conceive of movements originating from "eccentric" centers (Canguilhem 1955: 68, 127). Only in the late nineteenth century did physiologists place this concept of reflex within a mechanistic physiology. Thus Canguilhem can argue that concepts need not be homogeneous with the theories in which they ultimately will figure; concepts are "theoretically polyvalent" (*ibid.*, 6).

Althusser

Bachelard's epistemological oeuvre was not only read and used by historians of science like Canguilhem. Bachelardian notions and theses can also be found in the writings of the leading figure of French structuralist marxism, Louis Althusser. Perhaps it is no exaggeration to claim that at least part of Althusser's work can be read - and certainly was read so in the France of the late sixties - as an attempt to formulate, within a marxist framework, conclusions about the nature of science deriving in part from premisses that Althusser found in Bachelard and Canguilhem.

To begin with, Althusser rejects the notion, also repeatedly criticized by Bachelard and Canguilhem, that science finds its objects 'ready-made out there'. For Althusser, we have to free ourselves of such "mirror myths" by conceiving of knowledge as a production (Althusser 1970a: 24). In conceptualizing this production process, Althusser uses Marxian as well as Bachelardian terms. He suggests that this "theoretical production" always happens according to a specific, historically changing "*mode of production* of knowledges", i.e. "an *apparatus of thought*", or "a structure which combines (...) the type of object (raw material) on which it labors, the theoretical means of production available (its theory, its method and its technique, experimental or otherwise) and the historical relations (both theoretical, ideological and social) in which it produces" (*ibid.*, 41). Here "the raw material" is not a bunch of sense data or "the real object" out there, but rather the subject matter of science as it is constituted by the prevailing state of scientific knowledge, its technical means, and by relations between science and society (*ibid.*, 43; cf. Benton 1984: 36–39). Althusser also writes that this raw material, as object and problem, is what is "visible" for a given scientific field at a given moment in time. More precisely, this object is visible "within the horizon of a definite theoretical structure, its problematic, which constitutes its absolute and definite condition of possibility, and hence the absolute determination of *the forms in which all problems must be posed*, at any given moment in the science ..." (Althusser 1970a: 25). This field of the visible, however, is not opposed merely to what remains as yet unknown, but also – with reference to Foucault's works on madness and the clinic – also and primarily set apart from "the *inner darkness of exclusion*", i.e. from forms of knowledge that are ruled out and suppressed as unscientific within a given science (*ibid.*, 26–30).

In full accordance with Bachelard, Althusser proposes that there is no absolute standpoint from which the internal object of a science can be compared to the real object. Instead, theoretical practice itself defines its own criteria of knowledge and truth since it "contains in itself definite protocols with which to *validate* the quality of its product". The model here is the practice of mathematics:

> No mathematician in the world waits until physics has *verified* a theorem ... although whole areas of mathematics are applied in physics: the truth of his theorem is a hundred per cent provided by

criteria purely *internal* to the practice of mathematical proof ... (*ibid.*, 59).

Althusser also makes use of the Bachelardian notions of rupture and discontinuity. In contrast with Bachelard, he applies these concepts foremost to mark the difference between ideology and science. According to Althusser a new science can emerge only by rejecting "the pre-scientific *theoretical* ideologies that occupy the 'terrain' in which it is establishing itself" (1977: 13). Put differently, the prehistory of a science is always an "ideological theoretical practice" which is "qualitatively" distinct from, and discontinuous with, its history (*ibid.*, 168). Althusser is not very outspoken on what precisely defines an ideological theoretical practice but he does tell us that in this practice the validation of claims and theories does not yet happen by criteria internal to the theoretical practice, or that in this practice only such problems can be formulated whose "solution" has already been provided by "extra-theoretical instances and exigencies (by religious, ethical, political or other 'interests')" (1970a: 52). A further hint is provided where Althusser contrasts the scientific concept of socialism with the ideological concept of humanism: whereas the latter can only "designate" certain existing relations, the former "does ... provide us with a means of knowing them" (1977: 223). Perhaps this distinction between the ideological concept and the scientific concept is modelled upon the Canguilhemian opposition, encountered above with respect to reflex motion, between the description and the concept *qua* adequate interpretation of a phenomenon.

In *Reading Capital*, Althusser also suggests an "outline of a concept of historical time" (Althusser 1970b). Althusser seeks to show that Marx's *Capital* contains a theory of historical time which is able to improve upon the work of Braudel and other *Annalists*, and which is also of immediate relevance to the history of science. The ideological concept, from which this new theory has to radically separate itself, is the Hegelian concept of history and historical time. This concept is characterized by two main features, the "homogeneous continuity" and the "contemporaneity" of time. The first characteristic means that time is understood as a continuum "in which" the dialectical development of the Hegelian Idea through its various stages takes place. To conceive of time in this way, Althusser claims, makes it natural to see the defining and delimiting of historical periods as the central task of the historian. The second feature of Hegelian time, the contemporaneity of time, reinforces this tendency. The contemporaneity thesis is the assumption that all parts of a given whole, i.e. all elements of a given level of the development of the Hegelian Idea, coexist. None of them can lag behind the others, and none can run ahead of its time. In other words, this postulate assures the historian that wherever she makes a vertical cut through history, she will inevitably find "all the elements of the whole revealed by this cut ... in an immediate relationship with one another" (*ibid.*, 94).

Althusser claims that this Hegelian-ideological concept of time is still with us, and that not even Braudel, Febvre and serial history have fully broken with it. This is because even though they "*observe*" that there are different time spans and different temporal shapes, these historians have no theory on how these different times are interrelated. In the absence of such a theory, however, they have to fall back on the Hegelian notion of time and define their different times in terms of their duration, i.e. in terms of the "ideological time continuum" (*ibid.*, 96).

Opposing such a drawback, Althusser demands that the various temporal rhythms and shapes of different social practices and processes be interrelated not by the common measure of duration, but rather by a marxist theory of society that explains the specific nature or "relative autonomy" of each of these times, while ultimately, "in the last instance", explaining them in terms of the economic structure. Thus, the common measure of the various times that historians want to speak about is provided by one – or a bundle – of those times – namely the time(s) of the economy – rather than by an 'absolute time' like the Hegelian continuum (*ibid.*, 97).

Althusser speaks of economical time as the rhythms in which different modes of production develop their productive forces (*ibid.*, 99). This time is a specific time, yet "complex and non-linear":

> ... a time of times, a complex time that cannot be *read* in the continuity of the time of life or clocks, but has to be *constructed* out of the peculiar structures of production. ... The concept of this time must be constructed out of the reality of the different rhythms ...: out of the concepts of these different operations, e.g. the difference

between production time and labor time, the difference between the different cycles of production (the turnover of fixed capital, of circulating capital, of variable capital, monetary turnover, turnover of commercial capital and of financial capital, etc.) (*ibid.*, 101).

To construct the concept of a history of any domain, Althusser proposes, is to define the "historical facts", i.e. those events "which cause[..] a mutation in the existing structural relations" (*ibid.*, 102). For instance, in order to construct "the concept of the time of the history of philosophy" (*ibid.*, 101), we have to reject historical chronology, clarify the relation of philosophical practice to the sciences and ideology, and to structure the time of philosophy in terms of "*philosophical facts*" or "*philosophical events*" that "cause real mutations ... in ... the *existing theoretical problematic*" (*ibid.*, 102). This will lead us, Althusser believes, for instance, to regard Locke and Spinoza as more important than German Idealism, and to draw a direct line from Spinoza to Marx; after all, for Althusser, Spinoza is the only "direct ancestor" of Marx (*ibid.*, 102).

Althusser holds, furthermore, that his - or the Marxian - concept of time also passes beyond the continuity *vs.* discontinuity opposition as far as culture as a whole is concerned. Since his concept of time renounces the idea of the contemporaneity of time, i.e. since he rejects the idea that the elements of any time slice are all expressions of some core meaning like the Hegelian or Burckhardtian *Zeitgeist*, in any time slice there will always be both continuities and discontinuities: "The break 'valid' for a determinate level ... does not correspond to anything of the kind in the other levels ... which live in different times and know other breaks, other rhythms and other punctuations" (*ibid.*, 104). Likewise, Althusser is even skeptical about irregular or "stepped or multiply toothed cuts" or strata, since the construction of such strata is easily informed by the assumption that the elements standing out from the otherwise vertical surface are but further expressions of the one whole or core meaning (*ibid.*, 105).

Moreover, the notions of synchrony and diachrony can only be retained when understood along the lines of the new conception of historical time as construction. Synchrony is not the co-presence of two processes, events or things, but the existence of a construction of relations that connect these processes, events or things. *Mutatis mutandis*, the existence of diachrony between facts is nothing but the existence of a construction that defines transformations between them (*ibid.*, 108).

Finally, Althusser submits that the inability of historians to conceive of historical time as an interrelation of different times in the way suggested by his outline, is due to their lack of theories: "To put it crudely, history lives in the illusion that it can do without *theory* in the strong sense, without a theory of its object ... What acts as its theory is its *methodology* ..." (*ibid.*, 109). A methodology, however, that is not grounded in a theory of historical time – which in turn is built upon a theory of society – can only be ideological. In other words, as long as the discipline of history has no theory of its object, as long as it has no problematic, the discipline remains at the level of a prehistory of a scientific history still to come. It is bound to having its problems and objects determined by religious, ethical and political interests, and it is unable to define its own criteria of validation.

4. ARCHAEOLOGY, THE NEW HISTORIES, AND THE HISTORY OF IDEAS

It is one of the peculiar facts of twentieth century French intellectual history that the two new strands in historical scholarship, the *Annales* School and the epistemologists, have by and large ignored each other's work. Lucien Febvre's 1939 review of Bachelard's *Psychanalyse du feu* (Bachelard 1964) remained the only review of Bachelard's many books in the *Annales*, and none of Koyré's or Canguilhem's books was ever reviewed in this journal (Chartier 1982: 31). Furthermore, in a programmatic article on how the history of science should be written, "Sur Einstein et sur l'histoire" (Febvre 1955), published in the *Annales*, Febvre makes no mention of Koyré's, Bachelard's and Canguilhem's work. Febvre suggests that the study of the history of science be integrated into general history (*ibid.*, 306). Likewise, references to the works of the *Annales* historians are missing from the texts of the epistemologists.

The two schools of thought meet each other, it seems, only in the writings of authors that stand somewhat outside of both, i.e. in the writings of philosophers like Althusser and Foucault. Foucault's relations with both groupings are visible in that he published *Folie et déraison* in a series edited by Philippe Ariès, one of the leading figures of the *Annales* school, and *La naissance de la clinique* in a series edited by Canguilhem. As evidenced by favorable comments coming from Ariès, Braudel, Revel, Canguilhem and Althusser, Foucault also managed to gain recognition in both camps.[6]

Foucault's close awareness of developments in both schools is especially obvious in *The Archaeology of Knowledge*; in fact, large portions of this book can be read as a philosophical reflection of, and a critical comment upon, the state of the art in both areas of study.

Correlating the New Histories

In the introduction of *The Archaeology of Knowledge* (AK 3–17) Foucault is mainly concerned with stressing the points of convergence of *Annalist* and epistemologist paradigms. Indeed, he states as his aim, "to uncover the principles and consequences of an autochthonous transformation that is taking place in the field of historical knowledge" (AK 15). As Foucault sees it, these points of convergence are the concern with the definition of series, a new attitude towards the document, an attention to discontinuities, and a rejection of totalizations; in short, what unites both schools is the awareness that historical scholarship cannot but construct its subject matter.

While Foucault starts by depicting *Annalist* and epistemologist scholarship as somewhat diametrically opposed tendencies - concern with long periods in the one camp, eagerness to break up long periods into discontinuous strata in the other - he soon makes clear that both features are the natural result of the same basic constructivist trend. In both cases historians no longer stick to traditional, received periodizations and levels of events, but seek to define new trajectories in terms of new criteria. In the case of serial history, this leads to the isolation of ever new strata and series, to new criteria of periodizations and the search for novel interrelations between types of series (AK 4); in the case of the epistemologists it results in an attention to ruptures and transformations, in the breaking up of allegedly self-evident lines of tradition, and in the questioning of unities like the scientific discipline, the oeuvre of an author, the theory or the concept (AK 6).

Foucault characterizes the transformation, which issues in serial history and epistemological historical study in terms of the changed attitude towards the document. The document is no longer the primary unit of interpretation and of internal as well as external text-criticism. Instead, documents are organized into series, and broken up into parts and elements. Alluding to the archaeological practice of describing and analyzing monuments (i.e. archeological relics that, unlike artifacts, cannot be removed from the site of excavation, (Childe 1962: 11)), Foucault sets up an opposition between the new and the old histories in the following way:

> ... let us say that history, in its traditional form, undertook to 'memorize' the *monuments* of the past, transform them into *documents*, and lend speech to those traces which, in themselves, are often not verbal, ... in our time, history is that which transforms *documents* into *monuments*. ... in our time history aspires to the condition of archaeology, to the intrinsic description of the monument (AK 7).

Whereas the traditional historian took for granted the existence of certain 'natural' series and saw as her main task the placing of event-units within them, the modern historian first defines different types of events (e.g. according to their duration) and only then organizes the tokens of the respective types into series of different length and structure. Furthermore, whereas for the traditional historian "the discontinuous was both the given and the unthinkable", for the modern historian it is "one of the basic elements of historical analysis" (AK 8). For the traditional historian the discontinuous was given in the form of individual and unique events and documents. Subsequently, this discontinuity was transformed into continuity by drawing on totalizing concepts like historical evolution, *Zeitgeist* or *Weltbild*. In the hands of the *modernes*, however, discontinuity figures as operation, result and concept to be specified: it figures as "deliberate operation" in so far as the historian must start from the supposition that certain events can be organized into series in partial isolation from others; it is "the result of his description" in so far as the historian searches for ruptures in knowledge or points where cyclical changes start and end; and it is "the concept that the historian's work never ceases to specify" in so far as discontinuity is an umbrella concept: "one does not speak of the same discontinuity when describing an epistemological threshold, the point of reflexion in a population curve, or the replacement of one technique by another" (AK 9).

Taken together, Foucault concludes, the various ingredients of the conscious constructivism of the new histories amount to a new methodology of historical inquiry - and this despite the fact that many of the now-prevailing methodological problems have existed before. It also implies a deadly blow to the earlier project of "total history" and the philosophy of history by which it was informed. The various attempts of total history to secularize, by means of notions like *Zeitgeist* or *Weltanschauung*, the teleological Hegelian scheme turn out to be impossible projects once confronted with the serial historians' descriptions of ever new discontinuities, diversities, and asymmetries. The projects of total history must thus be replaced by a "general history" that confines itself to the description and explanation of ever more series in their partial interrelation (AK 9).

A Critical Re-Evaluation of the History of Ideas

Outlining what he takes to be the decisive turn of more recent French historical scholarship is only one of several strategies which Foucault employs in order to define the character of his archaeology of knowledge. Besides placing himself in the vicinity of serial history and epistemology, he also seeks to draw a line of separation between archaeology and a variety of other historical inquiries into knowledge and science. This variety of approaches is labelled, in *The Archaeology of Knowledge*, "history of ideas", even though the criticism that is levelled against the "history of ideas" is also, and by the same token, directed against certain viewpoints of the *Annalist* and epistemological schools, against assumptions prevailing in some quarters of the history of science, and against a number of philosophical approaches to history.

Rather than merely criticizing certain particular shortcomings of the history of ideas, Foucault wants to go further and establish that these shortcomings are grounded in the very character of this approach to history itself. Since this brand of historical research lacks a clearly defined object, its relation to other forms of historical studies remains obscure, its methods borrowed, and its criteria and categorizations unreflected. Moreover, it is based upon a pervasive evaluative scheme that it has inherited from the nineteenth century. To be sure, this is not to say that all studies carried out under the title "history of ideas" have produced nothing of value whatsoever. Rather it is to say, first, that the history of ideas lags behind many other forms of historical research in rigor and scholarship and, second, that its occasional misconstructions are not as coincidental as they might appear upon first sight.

In order to see that Foucault's analysis of the history of ideas is not unfair, we need to compare his claims with work actually done under this heading. As far as Foucault's more general claims are concerned, the natural reference point - at least for an English speaking audience - is of course Arthur Lovejoy. Lest it be suspected that other historians of thought have avoided the pitfalls of Lovejoy's work and program, I shall relate Foucault's central points to some other critics of the history of ideas, on the one hand, and to more recent studies in the history of ideas, on the other hand.

Foucault and Lovejoy

To begin with Foucault's claims that the history of ideas has "difficulties in demarcating [its] ... domain", "in defining the nature of its object", and in determining its relation to other disciplines (Foucault 1978: 18), it is not difficult to validate his assessment. For instance, in his programmatic paper "The Historiography of Ideas" (Lovejoy 1960: 1–13), Lovejoy starts from the observation that "historical study having to do, more or less, with ideas" can be found "under at least twelve different labels": the history of philosophy, the history of science, the history of language, the history of religious beliefs, literary history, comparative literature, the history of arts, economic history and the history of economic theory, the history of education, political and social history, as well as the historical part of sociology (*ibid.*, 1–2). Lovejoy wavers somewhat on whether the history of ideas *sui generis* is an interdisciplinary project involving all of these disciplines or whether it possesses a domain of its own. In the end, he comes down on the side of the latter option, arguing that it is the study of "unit-ideas" which distinguishes the history of ideas from the other disciplines mentioned. These unit-ideas, we are told, must be studied in relative separation from the twelve other disciplines because they can manifest themselves in all of the different domains of thought that those disciplines study:

> There is here another distinct realm of historiography, which needs to be added to the dozen mentioned at the outset, ... Until these units [i.e. unit-ideas] are first discriminated, until each of them which has played any large role in history is separately pursued through all the regions into which it has entered and in which it has exercised influence, any manifestation of it in a single region of intellectual history, or in an individual writer or writing, will, as a rule, be imperfectly understood ... (*ibid.*, 9).

Unit-ideas, as Lovejoy explains in the introduction to his famous *The Great Chain of Being* (Lovejoy 1973: 3–23), are a "decidedly limited" group of ideas that live through the centuries, combine in different ways in the works of philosophers, scientists, theologians and artists, are only rarely invented and die slowly (*ibid.*, 3–4). Unit-ideas can be "implicit or incompletely explicit assumptions or more or less unconscious mental habits" (*ibid.*, 7), "dialectical motives ... [e.g. a] turn of reasoning, [a] trick of logic, [a] methodological assumption" (*ibid.*, 10), "kinds of metaphysical pathos" (*ibid.*, 11), "sacred words and phrases", or "a single specific proposition or 'principle' expressly enunciated by the most influential of early European philosophers" (*ibid.*, 14).

Given this diversity one will certainly need to add even more disciplines to Lovejoy's bundle, for instance, disciplines like historical psychology, anthropology and the history of mentalities. Furthermore, and more importantly, note that Lovejoy manages to save the history of ideas from resolving into other disciplines only by means of two rather awkward assumptions: that a historical study of the surfacing of an idea in different domains can precede the study of the role of that idea within the respective fields, and that ideas can somehow be decontextualized. It is also worth recording that Lovejoy remains silent on the question as to the methodology by which such units are to be identified, or as to the criteria by which we decide when two texts express "the same" idea. To say that the history of ideas is "somewhat analogous to ... analytical chemistry" (*ibid.*, 3) is merely a metaphor, and Lovejoy's call for "philosophical analysis" remains somewhat empty as long as we are not told just what kind of philosophical analysis is involved (Lovejoy 1960: 8).

Foucault also claims that it is possible to identify "three great themes" within the history of ideas, to wit, "genesis, continuity, totalization" (AK 138): genesis in so far as the *origin* of ideas or themes is regarded as a crucial question; continuity and totalization in so far as different eras or different contemporaneous fields of thought are made homogeneous by the assumption of omnipresent elements.

Foucault regards these themes as guiding presuppositions in the choice of topics and the style of analysis of historians of ideas. With respect to the former, Foucault holds that the history of ideas favors the study of knowledge "which could never in the whole of its long, persistent life attain the form of scientificity" (AK 136). The history of ideas tries to find its niche between the histories of the sciences and literature by studying Diderot and Fontenelle rather than Descartes and Leibniz (cf. Foucault 1971a: 144), by attending less to "the great discursive monuments" than to widespread opinions, and by studying errors rather than truths, or mentalities rather than "forms of thought" (AK 137).

As concerns the typical style of investigation of the history of ideas, Foucault sees it as characterized by the dissolving of the works of historical figures into elements which are common to many fields or men in the respective period, on the one hand, and by a reconstruction of the life-histories (birth, gradual growth, maturity, diffusion, death) of themes, on the other hand (AK 137). The pervasive goals of the enterprise are to distinguish the original from the regular, and to reduce the multitude of differences or contradictions to an underlying harmony. First, the historian of ideas wants to determine who originally came up with a certain theme or idea or who combined prevailing ideas in the most unusual way (AK 141). Second, she is also eager to demonstrate that what at the surface seem to be crucial differences, can be traced back as different surface effects to an underlying, perhaps unconscious unit, or that diversity can be grouped around a few, or a single, oppositions (AK 149).

Again it seems hard not to concede to Foucault that his analysis is on the mark, at least as far as Lovejoy is concerned.[7] First, Lovejoy wants his followers to be "especially concerned with the manifestations of specific unit-ideas in the collective thought of large groups of persons" and to pay attention to common "beliefs, prejudices, pieties, tastes, aspirations, current among the educated classes through, it may be, a whole generation, or many generations" (Lovejoy 1973: 19). Even though he assumes that most unit-ideas can be traced back eventually to philosophers, he suggests that historians of ideas "seek for [their] most significant manifestations in art, and above all in literature" (*ibid.*, 17). Lovejoy also wants the history of ideas to lay bare the fact that the history of thought is "in great part, a history of confusions of ideas" (*ibid.*, 22). Second, Lovejoy reduces, in both synchronic and diachronic perspective, the diversity of philosophical systems, works of arts and theories of science to an antithetical conflict of a handful of unit-ideas. Moreover, he also calls for a sound knowledge of philosophy in order to be able to correctly find the philosopher-originators (Lovejoy 1960: 8) and he hopes to be able to "penetrate" behind "surface-dissimilarities" to "the common logical and pseudo-logical or affective ingredients" (Lovejoy 1973: 4). The urge to reduce plurality to harmonic simplicity is also evident where Lovejoy writes that the history of ideas "aims at interpretation and unification and seeks to correlate things which often are not on the surface connected" (*ibid.*, 21). Genesis, continuity and totalization thus figure indeed centrally in Lovejoy's program.

The mythology of ideas

Needless to say, Foucault does not stand alone in his criticism of the history of ideas.[8] Unfortunately, the amount of criticism levelled against the history of ideas, on the one hand, and certain of its practices with counterparts in other domains of historical scholarship, on the other hand, has not led to a situation in which this criticism has become obsolete.

Given his general view of the history of ideas, it is not surprising that Foucault directs his polemics against certain concepts and strategies which the historian of ideas (or of science, or of mentalities) uses in attempts to reduce differences, to operationalize her opposition between the original and the regular, and to vindicate some teleological scheme of progress.

To begin with the notion of tradition, Foucault suggests that the application of this concept is often marked by the preoccupation to reduce, in an arbitrary fashion, a variety of phenomena to the "form of the same" (AK 21). An unqualified use of this notion, i.e. a use that is not checked by the laying down of criteria for its use beforehand, makes it possible to relate almost any contemporaneous or noncontemporaneous elements. *The Archaeology of Knowledge* also proposes that the notion of tradition is linked to the search for origins and the category of the "genius": By means of an assumed tradition the historian of ideas can link ("at a distance and through time – as if through the mediation of a medium of propagation") a series of later individuals, notions or theories back to the founder as the ingenious originator of the chain to follow (AK 21).

Typically, the diachronic coherence of the sequence is established by reference to the slow and gradual evolution of one or several – typically ancient – ideas or themes, or by employing the concepts of influence, anticipation and precession. Like other critics of the history of ideas, Foucault is unsatisfied with all of these categories. First, the assumption of "unit-ideas", or "themata" leads away from the study of a specific discourse at a given time: this discourse is not "treated as and when it occurs" but referred back to a "semi-silence that precedes it" (AK 25). The employment of these notions and the interpretative strategies connected to them tacitly presupposes that all major ideas have always somehow been present below discourse until, finally, they "gradually come together and suddenly condense into the fine point of the work" (AK 138). Or, to put it in less metaphorical terms, for every important idea, it is assumed that it was "always in some sense immanent in history, even if various thinkers failed to 'hit upon' it, even if it 'dropped from sight' at various times, even if an entire era failed (note the implication that they *tried*) to 'rise to a consciousness' of it" (Skinner 1969: 10).

Second, the belief that every major idea has always been already immanently present - gradually evolving and maturing - also invites the notorious search for predecessors and anticipations. Every time some historian of ideas presents a claim of the form "X was the first to present idea Y", a chase starts to find anticipations, resulting in papers proposing an earlier instance of the same idea (Condren 1985: 101). The interpretative technique behind this procedure has been fittingly labelled "the emergence technique" and characterized in terms of two imperatives:

> Every idea has a predecessor which resembles it more closely than any other predecessor; find it! (Agassi 1963: 32).
> If a thinker's idea can be only partly traced to his predecessor, try to track the rest of his idea to another predecessor (*ibid.*, 34).

Thus historians have chased Bohr's planet system model back to the atom of Newton's alchemist writings, the Einsteinian concept of space to Aristotle, the Newtonian ether to the Stoics' *pneuma*, and the field theories of modern physics to Spinoza (Kragh 1987: 84–85, 99). As Foucault and others point out, however, to speak straightforwardly of the resemblance or identity between ideas or formulations, is to overlook that such resemblance and identity claims presuppose criteria which define "what is identity, partial or total, in the order of discourse" (AK 143).

Like Quentin Skinner, Foucault is also annoyed by the undifferentiated use of the category of influence. For both critics, this notion provides "a support - of too magical a kind to be very amenable to analysis - for the facts of transmission and communication" (AK 21). Both writers suggest that this notion should either be suspended altogether, or be applied only "in very precisely defined series" (AK 26, 143). Skinner's conditions are that the influenced writer could have got his doctrine only from the alleged source of influence, and that the probability of a coincidental similarity is very low (1969: 26).

Obviously, the search for predecessors, anticipations and influences is equivalent to the determination of degrees of originality: the "stock of originality" is raised or lowered by "degrees of nobility that are measured here by the absence of ancestors" (AK 143), and the significance of an author is evaluated in terms of the difference between the number of writers she has been influenced by and the number of writers influenced by her (Condren 1985: 131). For Foucault, these are "amusements for historians who refuse to grow up" (AK 144).

However, whereas Skinner sees the main genuine objective of the historian as being the identification of the "intention in writing" of authors (1969: 49), Foucault regards such a kind of goal as being part and parcel of the same bundle of ideas which characterizes the history of ideas. For him it is one of the central ingredients of this tradition to try "to rediscover beyond the statements themselves the intention of the speaking subject, his conscious activity, what he meant" (AK 27). In part this suspicion might be due to the observation that the endless debate over what this or that thinker "really" meant is an essential element of the evaluative game of lowering and raising the stock of originality, in part the suspicion is also motivated by Foucault's deeming it more important to study what in discourse becomes accessible to all or most writers of a given field and time, rather than to inquire what the specific intentions of a single writer in dealing with some topic are.[9]

The mythology at work: the examples of Holton and Cohen

Before going further and reviewing Foucault's criticism directed against concepts which figure as much in *Annales* and epistemological histories as in more traditional approaches, it might be useful to see that the arguments of Foucault and other critics of the history of ideas do still apply to the work done by leading historians of ideas and science of the eighties. For instance in the 1988 revised edition of his *Thematic Origins of Scientific Thought. Kepler to Einstein*, Gerald Holton restates his viewpoint according to which the study of "themata" provides a crucial key to the history of science. Like Lovejoy's unit-ideas, these themata can be catch-words, "quasi-aesthetic judgments", feelings, and principles. Furthermore, Holton's themata are long-living, appear jointly with one or two "antithemata", and "a total of fewer than fifty couples or triads seem historically to have sufficed for negotiating the great variety of discoveries" (1988: 17). Applying his theory, Holton suggests, among other things, that from the times of Milesian cosmology right down to the contemporary debates over evolutionary and steady-state cosmology, the same themata of evolution vs. constancy have been at work (*ibid.*, 45) and that the theme of discontinuity has been influential from antiquity to modern physics (*ibid.*, 17). In the light of such assumptions, one will no longer be surprised to find Holton express views like the following:

> ... at bottom, the work of major scientific 'revolutionaries', like Einstein, ... can usually be seen to be projections back to an idealized, purified state of the past - and, in particular, a return to a state of imagined classical purity, ... (*ibid.*, 93).

Not surprisingly, Holton makes no attempt to define criteria for the identity of themata in different authors, and he in no way answers the criticisms that have been levelled against the Lovejoyan program.

Since Holton's views have been criticized often enough, and since they might therefore be easily regarded as an exception, it seems more interesting to turn briefly to another leading figure in the historiography of science. The study I have in mind is I. Bernard Cohen's *The Newtonian Revolution* (1980). As the subtitle of the book, *With illustrations of the transformation of scientific ideas*, already suggests, Cohen argues that all creative scientific work in general, but also all revolutionary advances in particular, consist of "a series of transformations" of ideas. For instance, "Darwin transformed Lyell's concept of an interspecific competition or 'struggle for existence' to an intraspecific one, ..." (*ibid.*, 168).

Cohen proposes that an idea can either be transformed into a different idea, or then, like the notion of atom, "undergo successive transformations and continue to live on for a long time in science" (*ibid.*, 197). Furthermore, a series of transformations, starting from an idea, say p, can eventually lead to another idea, say q, which logically implies $\neg p$. Cohen speaks of the negative and positive "potentialities" of ideas, depending on whether a given idea is able "to serve a purpose quite opposite to any beliefs of an author" (negative potentiality) or not. For instance, the central ideas of Plato's *Timaeus* dialogue had negative potentiality since they led to the formation of a mathematical physical science based on experimentation and observation, an idea quite foreign to Plato himself. Along the same line, Cohen claims that different ideas have different potentialities of useful transformation, and that this usefulness is to be measured by the ideas' ability "to serve in the further advancement of science" (*ibid.*, 162). Moreover, the transformation of an idea can be conscious or unconscious, fortunate or unfortunate, correct or incorrect, and genius-dependent or genius-independent (*ibid.*, 168, 194, 195, 203, 221). The transformation is fortunate or unfortunate depending on whether or not the transformation leads to "good or useful or fruitful" ideas; the transformation is correct or incorrect depending on whether or not the transforming individual understands the original idea according to the intentions of its author; and the transformation is genius-dependent or independent depending on whether the transformation in question is "the last great step" of a revolutionary advance. Cohen writes that this last step "requires a mind of truly heroic creative proportions", a "supreme human genius" (*ibid.*, 162, 221). Finally, it deserves to be mentioned that Cohen regards Ernst Mach as his most important predecessor in stressing the importance of the notion of transformation. Mach held that ideas are gradually transformed "as in all likelihood one animal species is gradually transformed into a new species". Mach combined this gradualism with the conception of survival of the fittest: "Many ideas arise simultaneously. They fight the battle for existence not otherwise than do the Ichthysaurus, the Brahman, and the horse" (*ibid.*, 283–84).

Applying the criticism of the last section to Cohen, we can start from the observation that the biological-evolutionist metaphor is central in his approach. Cohen writes that historians and philosophers often treat "ideas as if ideas were persons" (*ibid.*, 327), but makes no critical comment on this style of writing history. He approvingly quotes Mach, who explicitly likens ideas to species. Cohen also writes that ideas "live on", have "existence", "survive", and "have potentialities" (*ibid.*, 197).

More importantly, in Cohen's talk of the potentialities of ideas it is not difficult to identify the tendency to read history backwards, i.e. to trace back a current "idea". Tacitly this procedure assumes that the current doctrine has always been in some sense immanent in history. That this is not an unfair accusation can be seen from the facts that Cohen wants to assess the fruitfulness of ideas and transformations in terms of the way they have advanced science, and that he allows for positive as well as negative potentiality: after all, the introduction of the negative potentiality of ideas makes it possible to say that earlier ideas are the seeds even of those later ideas which contradict them. Cohen's talk of the potentiality of ideas strongly suggests a vitalist's view of the *entelechy*, i.e. a power in the organism - here: in the idea - that directs the latter's development and completion. The Aristotelian notion of actuality, as resulting from a coincidence of *potentia activa* and *potentia passiva* also comes to mind because of Cohen's remark that a revolutionary advance presupposes the meeting of an idea "with potentialities for useful transformation ... with a mind capable of making the transformation" (*ibid.*, 168).

As concerns the "virus of the predecessor" - as Canguilhem liked to put it - Cohen seems to think that he can endorse this chase without falling prey to its implausible implications. Since the step from the predecessor to the successor involves a transformation of ideas, the historical dimension of science appears to be preserved (*ibid.*, 162–66). However, the credibility of this assurance is seriously undermined by Cohen's failure to provide us with a criterion for deciding whether a given transformation results in a new variant of an old idea, or whether the transformation results in a totally new idea. In one place, Cohen seems to suggest that the decisive criterion is the intention of the original creator of that idea which enters into the transformation (*ibid.*, 159). But this hint - which neatly supports the Foucauldian skepticism concerning intentions - clearly flies in the face of Cohen's example of the atom as one and the same idea despite successive transformations. It certainly seems absurd to say that Bohr remained truthful to the intentions of Democritos, and that thus Democritos can be regarded as a predecessor of Bohr's theory.

Finally, as concerns Cohen's interest in genius, it must be mentioned that he conceives of scientific revolutions as starting from the accumulation of various smaller-scale transformations, the resulting ideas of which are finally synthesized into a radical new set of ideas by a "supreme human genius" (*ibid.*, 221). As needs to be pointed out, this scheme is strongly reminiscent of the Lovejoy school's eagerness to distinguish between "radical originality" as the creation of new ideas, and "synthetic originality" as the creation of a new synthesis (see Condren 1985: 106). Since the number of ideas are regarded as decisively limited and long-living, the only way to increase their stock is by the implausible introduction of geniuses. Be this as it may, Cohen's preoccupation with genius and originality also shows, in any case, how central the evaluative game around the original vs. regular axis is to his framework.

The mythology of coherence and contradiction

The strategy of many historians of ideas and science to create coherence and downgrade differences comes to the fore not only in the evolutionary metaphor and the concepts linked to it. Foucault and Skinner also see it at work in certain interpretative strategies employed vis-à-vis texts, and Foucault identifies it moreover in concepts like *Weltanschauung, mentalité, Zeitgeist*, and in the science vs. non-science opposition proposed by Bachelard and Althusser.

Foucault's rejection of books and oeuvres as natural units is well in line with his general criticism of traditional document-based histories. Since the serial historian can accept units only in so far as she has first established the criteria for series and units, she cannot leave unquestioned the criteria by means of which texts or statements are grouped around the figure of the author. In the case of books, Foucault claims, it is not sufficient to base their coherence on "material unity", since this unity cannot account for the differences of, say, "an anthology of poems, a collection of posthumous fragments, Desargues' *Traité des Coniques*, or a volume of Michelet's *Histoire de France*" (AK 23). Moreover, a book is always caught up on a "network of references" to other texts, and in this network of statements the division in terms of book-units is only one of many.

The category of the oeuvre also needs to be critically examined. Criteria are needed for deciding why the historian includes or excludes, say, texts written under a pseudonym, various drafts of latter works, unpublished writings in general, letters, testaments, and what not. For instance, "the name 'Mallarmé' does not refer in the same way to his *thèmes* (translation exercises from French to English), his translations of Edgar Poe, his poems, and his replies to questionnaires" (AK 24). Put differently, the oeuvre of any author can be defined in several different ways, none of which is self-evident in itself. What is more, the focus on oeuvres in the history of philosophy, science and literature, is often informed by the tacit or open assumption that the whole production of a writer must be the expression of a single - or small set of - idea(s) (*ibid.*). Used in this way, an oeuvre, like already a book, can be the chief tool for reducing the number of differences and contradictions, that is, the main device for establishing coherence (AK 150). Thus the historian seeks to show that the different points of view of an author can be made to cohere by explaining them in terms of an underlying - explicit or implicit - theme or system; that the author's early, 'immature' works are simply so many attempts to formulate the views found in later ones; or that, in those texts which do not harmonize with his central doctrine, he was perhaps even 'hiding' his real views (Skinner 1969: 16–22; AK 150).

However, the quest for coherence is also operative on a more general level. It surfaces in the project of much of the traditional history of ideas to find below the different and contradictory phenomena of a given civilization, a common and unifying principle (AK 9, 22, 150). This is the project which we have already seen Foucault renounce as total history. The classic of this kind of writing history is of course Burckhardt's study of the Italian Renaissance, in which Burckhardt interpreted the Italian Renaissance as the age of the awakening individual. Foucault does not mention Burckhardt, but refers to the works of Cassirer, Goldmann, Lukács and Dilthey (Foucault 1969: 24; Foucault 1966). At various places, he also suggests that the notion of mentality can play a role similar to that of *Weltanschauung*, *vision du monde*, or *Geist* (AK 137, 209).[10]

In a review of the French translation of Ernst Cassirer's *Die Philosophie der Aufklärung* (Cassirer 1932), a book that promised to interpret the Enlightenment "in the unity of its intellectual origin and its determining principle ... out of One Central Point" (*ibid.*, vii–viii), Foucault argues that Cassirer's work exemplifies well the general tendency of "traditional histories", namely the tendency to reduce the multiplicity of forms of a given period first to the texts in which they were reflected at the time ("a theory of the world rather than positive science, ... an esthetics rather than the work of art, ... a philosophy rather than an institution"), and second to a philosophical principle common to these reflections (Foucault 1966: 4).

The same leaning, combined with a teleological perspective of progress and the strong preference given to the original over the regular, is of course pervasive in Goldmann. For instance, in his *The Human Sciences and Philosophy* (1969), Goldmann justifies the study of history as necessary for understanding the path from the past through the present towards "an authentic and universal community in the future" (*ibid.*, 29). The history of ideas is, for Goldmann, the study of the "consciousness of a class" (*ibid.*, 59), and he defines "the maximum of potential consciousness of a social class" as "a coherent world-view which may be expressed on the plane of religion, philosophy, literature or art" (*ibid.*, 103). Here potential consciousness is distinguished from "real consciousness" as "the totality of states of individual consciousness" and their interrelations (*ibid.*, 127). Potential consciousness is the collective consciousness that *would* prevail were it not for "multiple obstacles and deviations, ... the infinite variety and multiplicity of actions of other social groupings, and also of cosmic factors" (*ibid.*, 118). For Goldmann, it is the outstanding work of art, science and philosophy which is able to reach beyond the real to the possible consciousness. From this he infers that world-views enlarge due to the efforts of outstanding individuals, that we can best study the world-view of a period by attending to its outstanding works, and that these works provide us with the underlying coherence of real world-views (*ibid.*, 129–30).

Foucault claims that such schemes constitute a "*histoire pour philosophes*" rather than historical research proper. With respect to these schemes he also accepts the accusation made against him by Sartre, to wit, that he "is killing History" (Foucault 1968: 22).

The tendency to reduce contradictions and increase coherence can also lead to misconstructions in the (archaeologically and epistemologically orientated) study of science. In *The Archaeology of Knowledge* Foucault turns critically against *Folie et déraison* and *Naissance de la clinique* by observing that in these books he tried excessively to reduce the variety of objects in psychiatric discourse and the multiplicity of forms of statements in 19th century medicine to the singularity of one object, namely madness, in the first case, and one modus or style, namely description, in the latter case (AK 32–33). He now also feels more need to emphasize that the notorious *épistémè* of *Les Mots et les choses*, is not to be misunderstood as a unifying principle (AK 191).

As concerns epistemological science studies from Bachelard to Althusser, an exaggerated downplaying of differences can be seen at work. After all, the total break between common sense and science in Bachelard's case, or then the contradictory opposition between science and ideology in Althusser, are but versions of the traditional historian's project of either resolving various contradictions altogether, or reducing them until "the fundamental contradiction emerges ... as the founding, secret law that accounts for all minor contradictions" (AK 151).

Foucault argues, primarily against Bachelard, that in studying the emergence of a scientific discipline more than just one threshold must be distinguished. When in a given society a discourse with specific objects, styles and concepts has emerged, this discourse can be regarded as having passed the "threshold of positive discourse". This discourse need not yet be confined to any one institution (or clearly delimited group of institutions), for indeed at this level the discourse is typically spread out over many different domains and institutions. The discourse passes the "threshold of epistemology" once norms and models exist which allow the initially vague knowledge of the discourse to be checked. When the discourse has established rules as to how scientific knowledge is to be obtained, it has crossed the "threshold of science". And finally, once its theories are axiomatized, the scientific discipline has passed the "threshold of formalization" (AK 186–87).

Foucault suggests that different fields traverse these thresholds at different speeds, and that sometimes two or more thresholds can be passed at once. Disregarding these thresholds leads one to adopt "a model to be applied at all times and for all forms of knowledge: a division between what is definitively or what is not yet scientific" (AK 188). As *The Archaeology of Knowledge* proposes, this tendency to homogenize the development of sciences to one threshold is due to a mistaken assimilation of all sciences to mathematics; only in its case the four thresholds were indeed crossed at once. Yet precisely this fact makes mathematics a "bad example, an example at least from which one cannot generalize" (AK 189).

More specifically against Althusser, Foucault points out that ideology does not wear away once a problematic is established. First, as the examples of clinical discourse and political economy show, disciplines may well be deeply involved in the production or strengthening of ideologies and yet produce knowledge which is not throughout "undermined by error, contradiction, and a lack of objectivity". And second, by establishing more rigorous methods of proof and research, a discipline does not necessarily break with ideology: "The role of ideology does not diminish as rigor increases and error is dissipated" (AK 186).

5. THE ARCHAEOLOGICAL MODEL I: IDENTIFYING DISCURSIVE FORMATIONS

Introduction

Having reviewed where *The Archaeology of Knowledge* identifies friends and foes, we can turn to Foucault's own proposals on how the history of science might be written. Foucault's own, archaeological, model is most naturally reconstructed in two steps. I shall begin by introducing the main conceptual tools that Foucault employs for analyzing what he terms "discursive formations" and "statements". These notions will be related to the methodology of serial history, and sharpened by drawing on some basic notions from analytical philosophy. It is only subsequently that we can take up the *Gretchenfrage* for any philosophy or history of science, to wit, the question how continuities and discontinuities are conceptualized and accounted for. I hope to show that the archaeological model contains suggestions for describing change in science that are still of systematic interest today.

The negative, critical part of the archaeological enterprise is, as seen above, something like a 'deconstruction' of some of the allegedly 'natural' categories and units which historians of science and culture have traditionally employed. The positive, or constructive, side of *The Archaeology of Knowledge* consists of the introduction of a new set of concepts and questions. To put it in a nutshell, Foucault seeks to define, in a by and large systematic fashion, new kinds of events, series of these events, their parts as well as their conditions, and he wants to determine different series of series, i.e. different interrelations between series.

Now what remains, after we have gotten rid of the traditional units like tradition, influence, *Zeitgeist* and *vision du monde*, are simply aggregates of events. As Foucault sees it, these events, or fossils of the archaeological record, are "statements" (*énoncés*):

> ... the totality of all effective statements ... the material with which one is dealing is, in its raw, neutral state, a population of events in the space of discourse in general (AK 27).

To be exact, even statements are not something immediately given, of course, since looked upon without prior categories like book, oeuvre or tradition, the archaeological record, or "the discursive field" (AK 28), consists of just a mass of inscriptions. To speak of statements as events thus calls for a theory of what statements are, and how they are to be identified.

Relying for the moment on an intuitive understanding of 'statements', we might note, in any case, that to treat statements as events is not implausible; after all, acts of writing down or uttering statements are events, their printing or recording are events, and their repetition, translation, transformation, reactivation, or forgetting are also events. Furthermore, Foucault's speaking of the statements themselves as events too, can be justified by the parallel between statements and speech acts, a parallel that Foucault, after some initial hesitation, was ready to accept (Dreyfus & Rabinow 1983: 46). Since speech acts are – according to Austin's famous formula – acts of "doing things with words", they obviously are events in time.

While statements, their conditions, their parts and the objects to which they refer, form the basic, initial units of archaeological study, its series, or series of series, are "discursive formations" and their interrelations. As a first approximation, we can say that a discursive formation is simply a fairly large population of statements. In order to delimit this series, however, we need to identify several other series that ultimately mark out the conditions of existence of those statements. Put differently, to explain how a given series of statement-events is possible, demands that statement-events be analyzed in various directions, such as with respect to the objects to which they refer, the people who formulate them, the concepts they contain, and the different ways they can be combined. This investigation then leads to a more refined concept of discursive formations. From this more analytic perspective, discursive formations are practices displaying specific regularities, or, more precisely, discursive formations are practices constituted by the interrelation and correspondence between several series of factors and conditions.

In *The Archaeology of Knowledge*, Foucault initially leaves the statement undefined, and introduces discursive formations by way of examples from his earlier books. From the vantage point of my reconstruction, however, it seems more natural to start from the notion of the statement.

WHAT IS A FOUCAULDIAN STATEMENT?

A reconstruction of Foucault's notion of the statement is by no means an easy task. The 60-odd pages of *The Archeology of Knowledge* (AK 79–134) that elaborate on this concept are not only the most difficult part of the whole book, but they also contain some of the least transparent passages of Foucault's whole oeuvre. The reader's patience is challenged by definitions that are proposed only to be subsequently withdrawn, by a long list of what a statement is not supposed to be, and by conceptual distinctions that remain unclear to the end. Here I shall not try to discuss the twists and turns, nor try to criticize the obscurities of Foucault's text. Rather, I shall present my reading of what I take to be the hard – intelligible – core of Foucault's treatment of statements without going into every detail of his elaboration.

The fairly simple step of my interpretation is to take seriously Foucault's suggestion that dealing with statements means "dealing with an enunciative function" that relates signs "to a field of objects ... a number of possible subjective positions ... a domain of coordination and coexistence [and] ... a space in which they are used and repeated" (AK 106). The natural way of rendering this passage intelligible is, obviously, to take the notion of a function at its mathematical face value. Whether a string of signs is a statement or not, Foucault tells us, depends on its relation to other factors. What then is more inviting than to express this dependence with a function in its most general, set-theoretical sense? After all, functions in this sense are precisely the kind of things by means of which dependencies can be formally expressed. My proposal then is that a Foucauldian statement is an ordered quintuple

$$<r_i, p_i, a_i, l_i, s_i>$$

and that any set of statements ($\acute{E}$) can be represented by a function ($f_{\acute{E}}$) that maps various combinations of elements of sets R, P, A, L into elements of set S:

$$f_{\acute{E}}: R \times P \times A \times L \rightarrow S$$

Here R is the class of "referentials", P is the class of "subjective positions", A is the class of "associated domains", L is the class of limits of repeatability, and S is the class of strings or systems of signs. All of these five coordinates, or ingredients, of a statement call, of course, for some explanation.

(1) To begin with the notion of referentials, Foucault informs us that in order to qualify as a statement, a string of signs must "refer to something", i.e. must refer to objects and states of affairs. Each statement has as its "*correlate* ... a group of domains [e.g. of objects and states of affairs] a domain of material objects, ... a domain of fictitious objects ... a domain of spatial and geographical locations" (AK 91). This condition for statementhood is of course tantamount to the central idea of intensional semantics. In the latter, a proposition specifies a set of possible worlds in which it is true, viz. in intensional semantics a proposition specifies a set of possible worlds that contain states of affairs and objects to which the proposition refers. Similarly, a Foucauldian statement is related to a set of domains of states of affairs and objects: statements are correlated with domains in which they have reference.

There is, however, one additional complication in Foucault's notion of referentials that is not immediately captured by this parallel. Foucault draws a distinction between "correlate" and "referential" to the effect that whereas the correlate consists of the domains of objects, the referential consists of "laws of possibility" or "rules of existence" of these various domains (AK 91). Thus different domains are not simply pregiven. Instead, they themselves depend upon various factors and operations, they themselves are the result of a construction: what kinds of domains of objects are possible or accessible at a given time depends on a whole variety of relations between institutions, instruments, and systems of concepts. In order then to account for the differentiation between correlate and referential we might say that whereas the correlate of a statement consists of the domains in which it has reference, the referential comprises the conditions of the possibility of these domains.

(2) To turn to the class of subjective positions, *P*, Foucault writes that it is constitutive of a statement to prescribe "a particular, vacant place that in fact may be filled by different individuals ... To describe a formulation qua statement does not consist in analyzing the relations between the author and what he says ... but in determining what position can and must be occupied by any individual if he is to be the subject of it" (AK 96). In other words, each statement prescribes certain conditions that a speaker/writer must fulfill in order to qualify as the "author" of that statement.

In order to flesh out the *P* coordinate somewhat, it might be fruitful to take the *P* index as an expression of modality in the sense in which Latour and Woolgar use this term. According to these authors, statements in science can be classified by means of a fivefold distinction between modalities, these modalities being: (1) wild speculation, (2) plausible suggestion, (3) reporting the finding of others, (4) fact-stating, and (5) taking-for-granted. (Latour & Woolgar 1986: 76–82). Since in each of these cases the statement obviously prescribes different roles and positions for the speaker or writer, the parallel between these modalities and Foucault's subjective positions does not seem artificial.

(3) Each statement is always related to a domain of other statements that figure as its "associated domain", "collateral space", "background of a whole verbal network", "associated field", "enunciated field", or "enunciative network" (AK 96–99). No statement can ever appear in isolation, i.e., it is constitutive of a statement to be situated within, and to be determined by, a class of other statements. Although Foucault does not specify the upper limits of the associated field, he nevertheless makes it clear that the associated domain of a statement is much more than just an immediate context or co-text (AK 97). In fact, what counts as a possible context of some statement is determined by the rules and the structure of its associated domain. Thus, for instance, restrictions on context work differently in poetry and mathematics: whereas the two

statements "I love Ronny" and "2+2=4" can immediately follow each other in a poetic text, they can hardly do so in a mathematical treatise.

(4) In order to approach the class of limits of repeatability, *L*, I have to introduce some more of Foucault's terminological machinery. Especially his distinction between "statement", "enunciation", and "formulation" is important here. Both formulation and enunciation are bound to a single spatiotemporal location, but whereas a formulation is the act of stating a statement, an enunciation is merely the act of uttering a string of signs (AK 98, 101). Now, because an enunciation is merely an utterance of signs - and not yet an utterance of a statement - it allows for an abstraction over its spatiotemporal constants: "... this uniqueness [of the enunciation] allows for a number of constants - grammatical, semantical, logical - by which one can, by neutralizing the moment of enunciation and the coordinates that individualize it, recognize the general form of a sentence, a meaning, a proposition" (AK 101).

In the case of the statement, however, the type-token distinction works somewhat differently. Each formulation of a statement is a token of a type, but whether different formulations are of the same type or not, i.e. whether they are formulations of the same statement, is not determined by grammatical structure or logical form. Instead it is to be decided by studying the statement's "status as a thing or object", on the one hand, and by attending to its "field of stabilization", on the other hand (AK 103). To investigate the first is to study the limits and restrictions that a given culture, community or institution imposes upon the "possibilities of reinscription and transcription" of statements (AK 103). For instance, a psychoanalytically oriented literary critic might treat two formulations, one of which a novelist utters in an interview, and the other of which she places into the mouth of one of her characters, as constituting the same statement. A legal expert, however, who has to determine whether the formulations in question qualify as an insult against someone, might, in the same case, speak of two different statements of which only one can lead to legal action.

To investigate a statement's "field of stabilization", on the other hand, is to attend to the series of statements to which a given statement belongs. Here one seeks to determine whether two sentences which, when looked upon in isolation, seem to express the same proposition, do in fact refer to the same (kind of) objects, have the same (kind of) subject position, and belong to the same discursive series. Attending to such series, Foucault holds, forces us to say, for instance, that "the affirmation that the earth is round or that species evolve does not constitute the same statement before and after Copernicus, before and after Darwin ... [and that] the sentence 'dreams fulfil desires' ... is not the same statement in Plato and Freud" (AK 103).

(5) To turn, finally, to the sign coordinate of the statement, it suffices to mention that Foucauldian statements are not restricted to formulations in natural language; formalizations and graphical representations too can be statements (AK 86).

The Four Series

To delimit a population of statements as belonging to one and the same discursive formation is to identify a structure of interrelations holding between the conditions of existence of the statements' coordinates. Varying his terminology slightly, Foucault suggests that to study a discursive formation as to the conditions of its possibility is to identify the historical conditions of the possibility of the formation of its "objects", "enunciative modalities", "concepts" and "strategies". He emphasizes that discursive formations cannot be identified, and thus delimited from one another, by identifying for each of them a single central object, e.g., madness, by identifying for each a single common mode of language or "enunciative modality", e.g., descriptive or prescriptive language, by identifying for each a clearly confined small set of concepts, or by identifying for each a single theory or theme. Instead, a discursive formation is identified once a law-like regularity of dependence or interrelation can be detected, a regularity that expresses the dependencies and interrelations between conditions of the possibility for a variety of objects, a variety of enunciative modalities, a variety of concepts, and a variety of theories or themes (AK 32–35).

These four coordinates of the discursive formation do not correspond exactly to the five coordinates of the statement, but as we shall see this asymmetry has little bearing upon substantial issues. It is more important to turn to Foucault's proposals as to how each of these coordinates are to be studied.

Objects

As concerns the constitution of *objects*, *The Archaeology of Knowledge* submits that we attend to three factors: "surfaces of emergence", "authorities of delimitation", and "grids of specification" (AK 41–44). Surfaces of emergence are the places and spaces, i.e. institutions or cultural fields, in which certain elements emerge as objects for discourse. In the case of the rise of psychopathological discourse in the nineteenth century, Foucault claims, such surfaces were, among others, the family, the immediate social groups, the working place, the religious community and art (AK 41). It was in these social and cultural contexts that madness was first systematically looked for and - alas - found. Authorities of delimitation were those individuals, groups and disciplines that came to be regarded as qualified for the identification of these new objects, foremost doctors, judges, art critics and priests. Finally, grids of specification were the systems of concepts "according to which the different 'kinds of madness' are divided, contrasted, related, classified, [and] derived from one another as objects of psychiatric discourse" (AK 42).

Or, to take as another example Steven Shapin's and Simon Schaffer's study on the emergence of experimentalism in natural philosophy of mid-seventeenth century England, we might say that the surfaces of emergence of a new scientific object like Boyle's "spring of the air" were the laboratory and the assembly room of the Royal Society with its crucial instrument, the air-pump. Authorities of delimitation were Boyle and his followers, the "physiologists", who severely restricted access to this new space, who evaluated and judged the credibility of observations reported, and who determined who was to publish in their *Philosophical Transactions*. Their grid of specification consisted not only of the traditional corpuscular and mechanical notions, but also of new concepts relating to different aspects of the spring of air. Their grid was also marked by a tight distinction between the language of theory and the language of facts (Shapin & Schaffer 1985: 38–39, 45, 57–58, 71,

135–136, 180–181).

What Foucault wishes to claim is that the objects of a (scientific) discourse are not waiting 'out there', but are rather the result of a group of relations that exist within and between surfaces, authorities and grids. In other words, "it is not enough for us to open our eyes, to pay attention" in order to encounter new scientific objects, rather these new objects exist only "under the positive conditions of a complex group of relations". Furthermore, these relations hold "between institutions, economic and social processes, behavioral patterns, systems of norms, techniques, types of classification, modes of characterization; and these relations are not present in the object ..." (AK 45). Foucault calls these object-constituting relations "discursive relations" and distinguishes them from "primary" and "reflexive relations". Primary relations are the relations between, say, the doctor and the family, relations that can be described without attending to the doctors' medical discourse. However, both reflexive and discursive relations are relations pertaining to the doctors' purported knowledge. Reflexive relations are relations that figure explicitly in the doctors' discourse, are those relations that he describes and analyzes. (For instance, the relations between family life and criminality.) Discursive relations differ from reflexive relations in not being formulated in the doctors' discourse: discursive relations make certain objects of the doctors' discourse possible, constitute objects (like madness) that the doctors come to identify, amongst other places, in the families (AK 45).

The isolating of discursive relations is crucial to the archaeological enterprise. To see why this is so we need to recall the central problem of serial history, to wit, the delimitation of series. *Prima facie*, the wealth of different objects of different discourses, and the continuous emergence of new ones, might make it seem impossible to pick out a group of objects as forming one distinct series. Foucault proposes, however, that this pessimism turns out to be unfounded once we attend to discursive relations. According to his suggestion, we identify a series of objects once we have shown for a group of non-contemporaneous objects that they emerge on the basis of the same set of discursive relations; in other words, that they have the same (social, institutional and conceptual) conditions of possibility:

> ... a discursive formation is defined (as far as its objects are concerned, at least) ... if one can show how any particular object of discourse finds in it its place and law of emergence; if one can show that it may give birth simultaneously or successively to mutually exclusive objects, without having to modify itself (AK 44).

Enunciative modalities

Turning to the second series that archaeology seeks to define, the "subjective positions", or "enunciative modalities", we encounter the same principle of identification albeit now defined in terms of three different factors. The first of these factors might be called 'speaker's position' and includes criteria of competence, the relations of the competent speakers to other social groups and the characteristics that define their status in society (AK 50–51). In the case of mid-seventeenth century 'physiology' an analysis of this element would have to include a description of the following norms: in order to qualify as competent members, researchers have to communicate a candidate matter of fact, have to accept experiments as the touchstone of natural philosophy, have to reject metaphysics and have to abide by specific rules of criticizing and writing. One would also have to mention the role of experimentalists in Restoration England, as well as their self-conscious posing as social arbitrators and as "priests of nature". These latter roles are important for understanding the support Boyleans received from the Church and the State (Shapin & Schaffer 1985: 70, 174, 184, 216, 319).

To comprehend the conditions of a given group of enunciative modalities also calls for a study of "institutional sites" and "positions of the subject". The former refer to places from which the competent subject makes his discourse (e.g. the assembly hall, the laboratory, the book written in English or Latin), the latter to his or her position in the information network and to the different roles the subject takes up in relation to his or her objects (AK 52–53). In the case of the "physiologists", such typical postures were that of the eye-witness, of the faithful reporter, of the modest narrator of successes and failures, and of the receiver and writer of letters circulating within a small group of natural philosophers, theologians and philosophers (Shapin & Schaffer 1985: 56,

59, 60, 68, 319).

Organizations of the field of statements

Under the title of a study of "the formation of concepts", Foucault provides a longish list of categories for the investigation into various relations between statements. This is because *The Archaeology of Knowledge* deems a study of statements an indispensable groundwork for the study of concepts. Since a detailed scrutiny of this list lies somewhat to the side of our main interest, suffice it here to summarize these distinctions by way of a table (AK 57–68). (Figure 1)

For instance, the discourse of the English physiologists showed the following "organization of the field of statements": its statements were *ordered* according to the logic of description, i.e. the narrative and the letter. Demonstrative reasoning and the form of the dialogue were rejected as orderings of the dogmatic philosopher. The central *type of dependence* was the hypothesis / verification dyad, and the typical *rhetorical scheme* was the essay, as "the piecemeal reporting of the experimental trials". The architecture of Boyle's text was also marked by the "conspicuous interval" that Boyle wished to leave between his narratives of experimental findings and his "discourses" on their interpretation (Shapin & Schaffer 1985: 39, 40, 44, 66, 67, 143). Concerning forms of coexistence of statements, the *field of presence* of Boyleans contained the philosophical discourse on vacuism and plenism only as something to be rejected and transformed into the language of experiments. An intriguing example of a Boylean *field of concomitance* was the legal discourse on eye witnessing, a discourse which was applied *per analogiam* to the witnessing of experiments. The *field of memory* of the new experimental science contained philosophical-metaphysical demonstrations, on the one hand, and some references to the alchemists, on the other hand. The statements of the former were those in relation to which Boyle established a relation of discontinuity, whereas the latter figured occasionally as predecessors of his own enterprise (*ibid.*, 45, 56, 71). Finally, with respect to characteristic *procedures of intervention*, one should at least mention the use of tables, and the central role of geometry rather than arithmetic as an important measure for *transcribing*

FORMS OF SUCCESSION

- *ordering of statement series*
  e.g. demonstrative reasoning
  descriptions
- *types of dependence*
  e.g. hypothesis / verification
  assertion / critique
  law / application
- *rhetorical schemata*
  the architecture of the text

FORMS OF COEXISTENCE

- *field of presence*
  statements taken up in a given discourse positively or negatively; e.g. as correct description, good reasoning, necessary presupposition, as authoritative statement ...
- *field of concomitance*
  statements of other discourses taken up as analogical confirmation, general principle, model, higher authority ...
- *field of memory*
  statements no longer accepted as valid, but in relation to which relations of filiation, continuity and discontinuity can be defined

PROCEDURES OF REWRITING

- *techniques of rewriting*
  e.g. linear descriptions in tables
- *method of transcribing statements from natural into formal languages*
- *modes of translating quantitative into qualitative statements and vice versa*
- *means of increasing the approximation of statements, to extend and restrict their validity*
- *ways of transferring a statement to a new field of application*
- *methods of systematizing statements*

Figure 1

statements from natural language into another sign-system (Shapin 1988: 23–58).

Strategies

The fourth kind of series that Foucauldian archaeology seeks to define is the series of "strategies". Under this heading *The Archaeology of Knowledge* groups theories and themes and proposes that in their case too we can define certain regularities of conditions of possibility. Defining these regularities involves the identification of "possible points of diffraction of discourse", of "the economy of the discursive constellation" and of the discourse's "function in a field of non-discursive practices". Points of diffractions are points where a given discourse branches off into sub-discourses: these junctures are "points of equivalence" in the sense that the two or more alternatives are determined by the same regularities concerning the conditions of objects, concepts and modalities; they are "points of incompatibility" in the sense that at the juncture a choice has to be made; and, they are "link points of systematization" in that both alternatives can give rise to a new series of statements (AK 65–66). For instance, Hobbes's interpretation of some of the results obtained in Boyle's air-pump experiments meet these conditions. Phenomena which Boyle regarded as clear proofs of the existence of the spring of air, like the dying of animals in the allegedly 'evacuated' receiver, Hobbes interpreted as effects of the uneven mixtures of extremely small, earthy particles moving rapidly in a circular motion (Shapin & Schaffer 1985: 122).

The study of the conditions of theories is also the study of the conditions for, and the regularities of, the mechanisms which close disagreement and choice. These mechanisms of closure are the "economy of the discursive constellation" and the "function in a field of non-discursive practices". The first factor is meant to capture the relation of a given (scientific) discourse to other, contemporaneous discourses which function as model, analogue, opposite or complement. For example, in the case of the seventeenth century physiologists, theory-choice was influenced by the partial rejection of the alchemists, and the total rejection of metaphysics. Theology, however, was seen as an ally (*ibid.*,

139). The second factor comprises elements like the role of the respective discipline in non-scientific practices, e.g. in education or manufacturing, the intervention of non-scientific institutions into science, and "the possible positions of desire in relation to discourse" (AK 68). For example, Boyle's victory over Hobbes was due in part to the fact that priests as well as lawyers were opposed to Hobbes. In order to translate the theologians' concerns into support for his theories, Boyle justified his nescience as to real causes by the theological argument that God can produce the same effect by a number of different causes. As far as desire was concerned, be it mentioned that the air-pump was repeatedly used for the entertainment of royalty and politicians, and that - at least in iconography - the air-pump promised an approximation of God's knowledge and the reaching for the invisible (Shapin & Schaffer 1985: 30, 32, 36, 37, 153, 313).

Interrelations, Derivation Trees and Contradictions

Defining series of objects in terms of their common conditions of possibility, and doing likewise for enunciative modalities, concepts, and strategies, does not yet suffice for identifying a discursive formation. In order to isolate a discursive formation, *The Archaeology of Knowledge* claims, we have to go further and study correlations and interrelations between the four kinds of series. Foucault also writes that a discursive formation is "a scheme of the correspondence between several temporal series" (AK 74). I take this to mean that of the four temporal series defined each has its own rhythm of change and its own specific limits, its own duration. In other words, for us to speak of a discursive formation, it is not necessary that the limits of all of the four series coincide in time. What is called for, as a justification for calling a given group of statements a discursive formation, is rather that the four series of objects, modalities, modes of organization of statements and strategies coincide or overlap for some time, and that we can establish interrelations between these series during this time (Figure 2). (Writing "*O*" for a series of objects, "*EM*" for a series of enunciative modalities, "*C*" for a series of concepts and organizations of statements, "*T*" for a series of theories and strategies, "↕" for interdependence between given series).

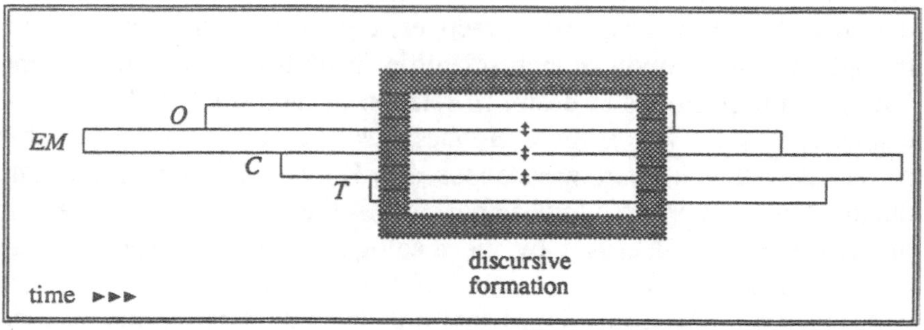

Figure 2

To quote Foucault more fully:

> A discursive formation ... presents the principle of articulation between a series of discursive events and other series of events, transformations, mutations, and processes. It is not an atemporal form, but a scheme of correspondence between several temporal series (AK 74).

Foucault does not provide us with a list of possibilities as to how these strata can influence or determine each other, but he stresses, in any case, that the direction of conditioning, restricting and delimiting does not necessarily take a uniform direction from upper strata to lower ones, or from lower to upper ones. These interrelations are rather of various kinds, and can change during the existence-span of the discursive formation (AK 73–75).

The last mentioned point is worth underlining, since it shows that a discursive formation is not a static set of fixed regularities. This point should already be clear in so far as the four series defined are series of simultaneous as well as successive, even in part mutually exclusive, objects, modalities, concepts and theories. To this we must add that the interplay of the four series can even involve the altering of some local regularity, (e.g. of a regularity holding within the conditions of the emergence of some objects), without, however, the discursive formation being itself changed fundamentally. In fact, the existence of such kinds of influences is as good evidence as any that one is in fact dealing with a discursive formation (*qua* interrelation between various series). To cite Foucault's own example:

> The hospital field ... did not remain unaffected when clinical discourse was put into relation with the laboratory: the body of rules that governed its working, the status accorded to the hospital doctor, the function of his observation, the level of analysis that can be carried out in it, were necessarily modified (AK 74).

Allowing for changes on the level of local regularities within one and the same discursive formation obviously presupposes a distinction between regularities of different generality. And indeed, in several places Foucault distinguishes between more fundamental or "general rules" and "more specific" or "derived" rules (AK 147, 168, 177). This distinction can be criticized as far as this talk of *rules* (here as elsewhere in the archaeological edifice) is somewhat misplaced; after all, if rules are to be more than regularities they must somehow be internalized by someone. Nevertheless, the distinction between different levels of regularities is certainly called for and helpful, especially when we turn to discontinuities and revolutions in science. Since this is a topic to be taken up further below, suffice it here to mention two other uses to which Foucault puts this distinction.

The first case concerns what Foucault calls the construction of "the tree of derivation" of a discursive formation. *The Archaeology of Knowledge* is not especially clear on how this tree is to be drawn, but it is plain that at the roots of this tree one places such statements that display the most general regularities of the discursive formation in question. These "guiding statements" (AK 147) do not *formulate* the discursive relations, since these are, as earlier heard, inaccessible for the discourse they regulate. Instead, guiding statements might be looked upon as counterparts of discursive relations within discourse itself. Guiding statements do not *formulate* but *mirror* discursive relations; the way they refer to objects, use concepts and form theories displays, on a general level, the outer limits of what it is possible to speak about in a given discursive formation. Or, put differently, guiding statements formulate in the forms of general norms, presuppositions, axioms or exemplars the limits of discourse in so far as these limits are accessible to discourse itself. (Perhaps we can say that Boyle's statements on how experimentalists are to go about doing science and how they are to achieve agreement are in this sense guiding statements of seventeenth century experimentalist science. The social and institutional conditions of the possibility of this experimental form of life are not directly, or only distortedly, present in his discourse, but his programmatic statements make them easy to identify.)

Derived statements, on the other hand, are further up on the branched tree structure and display more local regularities. They are less fundamental than guiding statements in the sense that both their acceptance and rejection is consistent with the guiding statements, while they themselves are consistent only with the acceptance of the guiding statements.[11]

The relation between guiding and less fundamental statements need not be one of temporal succession, however. Foucault also allows for the possibility that the guiding statements of a given discursive formation are formulated only late within the existence-span of a discursive formation: "For certain discursive formations, the archaeological order [of derivation] is perhaps not very different from the systematic order, as in other cases it may follow the thread of chronological successions" (AK 148).

Only once we understand the structure of a given discursive formation in this way does it also become possible to distinguish between different types of contradictions or disagreements. "Derived contradictions" concern branchings fairly high up in the tree-structure, "intrinsic contradictions" are located in mid-air, above the roots and below the finer branches, and "extrinsic contradictions" are contradictions between statements that belong to two different discursive formations (AK 154). To take a more contemporary example, we might say that we encounter a derived contradiction where two linguists, both working within the framework of Chomsky's Theory of Government and Binding, disagree over the question whether or not to allow for rightward movements of elements in the process of transformation; we are dealing with an intrinsic contradiction in the case of two linguists that take different sides on whether or not to accept transformations within a generative framework (e.g. Gazdar's Generalized Phrase Structure Grammar *versus*

Chomsky's GB Theory); and we have an instance of a still more fundamental difference, perhaps an extrinsic one, where two linguists disagree over whether the study of language should take its starting point from Chomsky-style syntactical theories or the ethnomethodological description of conversations.[12]

Counterparts and Accessibility Relations
Comparing Discursive Formations

Having reviewed the central conditions or "rules" by means of which archaeology delimits discursive formations, we can turn to what Foucault calls "the comparative facts" (AK 157–65), i.e. relations of similarity and difference between discursive formations. Foucault's suggestions concerning these topics deserve attention, since in recent years several historians and philosophers of science have called for theories and historical studies on relations between disciplines.[13]

Discourses and discursive formations

Above I pointed out that a discursive formation can be characterized either by listing the statements of which it is composed, or by describing the system of interrelations between the four series, the system which constitutes the condition of the possibility of these statements. In other words, even though discursive formations are made up of statements, not every statement can be part of just any discursive formation. Foucault expresses this conception by writing that each discursive formation is characterized by a specific "law of coexistence", "principle of rarification", or "law of enunciative poverty" that restricts the discursive formation to precisely the statements it in fact is constructed to have (AK 116–20).

When couched in this way, it is inviting to suggest that what appears as "enunciative poverty" from one perspective, might just as well be regarded as "enunciative plenitude" from a different angle. While we perhaps cannot say that every statement which is possible with respect to the regularities of a given discursive formation *must be actual* sooner or later, we can say, in any case, that every statement that is possible, according to the laws of the discursive formation in question, can only become actual within this same discursive formation.

This interpretation also invites the assumption that there can be no identity of statements across the borders of a discursive formation. First of all, Foucault proposes that the statement is determined in its identity by "conditions and limits ... that are imposed by all other statements among which it figures" and that identity of wording is not enough to safeguard identity on the statement level. Thus, for example, the statement that the earth is round, is a different statement in the discursive formations before and after Copernicus (AK 103). In other words, a statement is repeatable only within one and the same discursive formation. 'Transdiscursive identity' is ruled out.

However, lest discursive formations begin to look like closed Leibnizian possible worlds, it cannot be stressed too much that this denial of transdiscursive identity holds only for statements, not for objects, modalities, concepts or even theories. This point does not always come out very clearly in *The Archaeology of Knowledge*, and deserves some closer scrutiny.

As Foucault has it, statements are grouped into discursive formations on the basis of whether or not their coordinates are restricted, or made possible, by one and the same system of interrelations between the four series. In other words, it is not enough for two statements to refer to the same objects in order for them to belong to the same discursive formation. It is also necessary, first, that their other coordinates are elements of the same series as well, and, second, that the four series from which both statements draw their coordinates, are interrelated according to one and the same system of interrelations.

Now while there is nothing implausible about this suggestion as such, it should nevertheless be made explicit that it must also leave room for statements that do not belong to any discursive formation. If we accept the tight Foucauldian conditions that a set of statements has to satisfy in order to qualify as a discursive formation, then we obviously must leave room for many discourses which are unstructured by archaeological standards (which of course does not mean that in such discourses "anything goes"). This observation suggests that discourse should be defined as a broader concept than discursive formations, and that discursive formations are best regarded as discourses of a special kind (as defined by Foucault and reconstructed above). One is then led to say that there can be statements outside of discursive formations, but not outside of discourses.[14]

Since for Foucault a discursive formation only exists once all of its characteristic series are present in a characteristic interplay, it also seems natural to assume that two different successive formations will often or typically be separated by a discursive phase in which such characteristic interplay, such a system of interrelations, does not exist. Of course to assume such a less structured phase is not the same as claiming that both discursive formations form *one* historical line of tradition or discipline and that the intermediate phase, or "feature interface", is one of a Kuhnian crisis.

Finally, as the example of Foucault's own historical studies on the emergence of different discursive formations suggests, it is often useful to describe the statements of an intermediate discourse as "fuzzy" elements either of prior or of subsequent discursive formations. According to this suggestion, a fuzzy element of a discursive formation is an element of an intermediate discourse which displays the regularities of the respective discursive formations only in part. Elements of the intermediate discourse are then not totally set apart from full members of a prior or later discursive formation. (Using, for the purpose of illustration, numerical values, we might say that full members of a discursive formation are assigned the value *1*, full members of a successor formation assigned the value *0*, and statements of the intermediate discourse a value α, such that $1 < \alpha \leq 0$. (Figure 3))

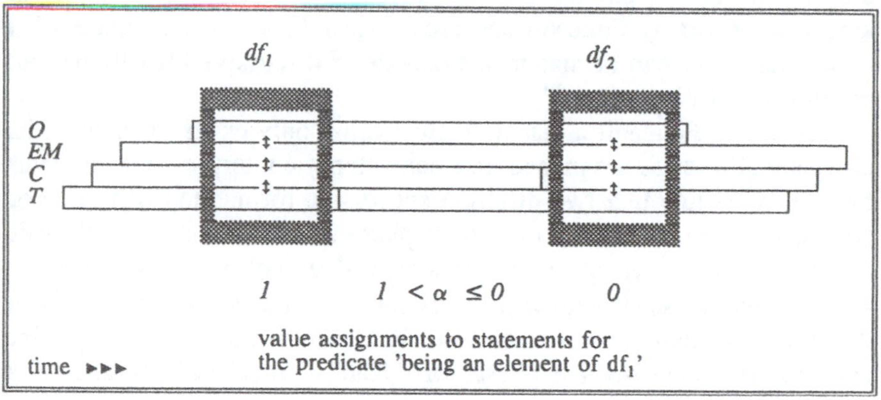

Figure 3

Counterparts

Although Foucault rejects the notion of transdiscursive identity of statements, he does, nevertheless, allow for transdiscursive *similarity*. My employment of the notion "statement-counterpart" should thus be clear enough. After all, it is due to the writings of David Lewis that the idea of transworld similarity has gained currency under the label of "the counterpart theory" (e.g. Lewis 1983a). Simply put, a counterpart of some individual, say *i*, of a given world, *v*, is the counterpart in *v* of another individual, *j*, in another world, *w*, if and only if, *i* resembles *j* more than any other individual of *v*.

The naturalness of transferring Lewis's notion to historical inquiry - even quite apart from Foucault - can also be defended by briefly indicating that Lewis's characterization of this relation as not transitive, not symmetric, and not a one-to-one relation is also intuitively convincing for the relation between statements. Let *1* to *9* be statements of three discursive formations, *A*, *B*, and *C*, as indicated in Figure 4.

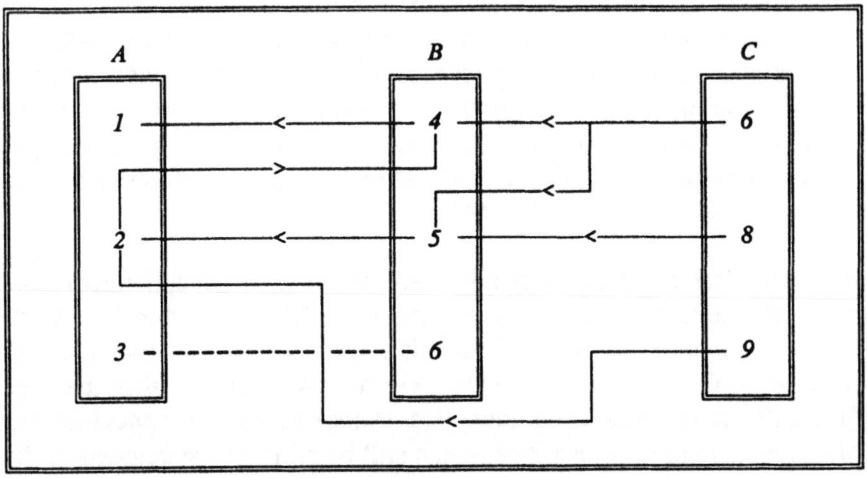

Figure 4

The counterpart relation is not transitive for even though *5* might be more like *2* than any other statement of *B*, and even though *8* is more like *5* than any other element of *C*, it still does not follow that *8* is more like 2 than any other element of *C*. In fact, it might be *9* that is more like *2* than any other element of *C*. Nor is the counterpart relation symmetric: *4* might resemble both *1* and 2 very strongly, more than any other statement of *B*. Thus *4* is a counterpart of *1*. Yet it might be that 2 resembles *4* more than *1*. Thus the counterpart of *4* in *A* is *2* and not *1*. Furthermore, the counterpart relation is not a one-to-one relation: *4* and *5* might resemble *7* more than anything else in *B* and they might resemble it to equal degrees. Thus they are both counterparts of *7*. Obviously, it is then also natural to allow that one statement is the counterpart of two statements. Finally, *3* might resemble *6* more than anything else in *A*, but still not be very like *6* at all. Then *3* is no counterpart of *6*.

In Foucault, the main instances of what I have been calling discursive counterparts are "archeological isotopia":

> [They are] entirely different concepts (like those of value and specific character [in the discursive formations of 17th century economics and of natural history respectively] ...) [that] occupy a similar position in the ramification of their system of positivity [i.e., discursive formation] – although their domain of application, their degree of formalization, and above all, their historical genesis makes them quite alien to one another (AK 161).

It is true that Foucault is talking here about concept counterpart, not statement counterparts. However, the transition from the first to the second can be justified by Foucault's pronouncements that concepts function only within statements. Furthermore, in another passage, Foucault argues that even though a statement cannot appear in two different discursive formations, it can still be related to statements within another discursive formation that are either "linguistically analogous", i.e., that have the same syntactic-lexical surface structure, or that are "logically identical", i.e., that are equivalent (AK 145). Thus altogether, we are provided with three criteria of statement similarity across discursive borders.

Accessibility relations

It also seems useful to maintain some kind of parallel between Foucault and Lewis as far as the relations between discursive formations are concerned. In Lewis's possible worlds conception every possible world is surrounded by a sphere of overall similar worlds. This notion of similarity of worlds is needed to account for the semantics of counterfactual sentences. For instance, to say that 'If Saddam Hussein were an American, Saddam Hussein and George Bush would be compatriots' is to say that there is a set (sphere of accessibility) of possible worlds, such that they are all similar to our world and such that Saddam Hussein – or his counterparts – figure as American citizens (Lewis 1973: 8–9).

Foucault, in turn, is interested in the question as to what conditions are required for two or more discursive formations to form a "region of interpositivity" or an "interdiscursive group" (AK 159). This question subsequently leads to the problem as to what conditions are needed for two or more discursive formations to be accessible to one another. In other words, Foucault holds that communication between discursive formations is possible only when the discursive formations in question are similar in some respects. Contrary to Lewis, in whose *Counterfactuals* the notion of similarity between possible worlds remains a primitive, unspecified concept, Foucault provides us with three criteria for similarity between discursive formations. We have already referred to the first one, to wit, the existence of "archeological isotopia", i.e., concepts or statements with a similar function. The second and third criteria are "archeological isomorphisms" and "archeological models". Foucault introduces these notions somewhat telegraphically, but it would seem that what he is trying to capture by means of these notions are certain parallels on the level of discursive relations. Even though two discourses might deal with different objects, have different enunciative modalities, different concepts and different theories, it is still possible that both owe their respective existences to similar conditions and regularities. Foucault speaks of "archaeological isomorphisms" when we find such common conditions in two discursive formations, and of the same "archaeological model" when even the rhythms of internal development and change are similar. (Foucault holds that Natural History, the Analysis of Wealth and General Grammar of the latter half of the seventeenth and of the eighteenth centuries show archaeological isomorphisms but have dissimilar archaeological models (AK 160–61).)

Now, since similarity between discursive formations is the condition for the possibility of their "law of communication", which in turn is the condition for the possibility of "archaeological correlations" as "relations of subordination or complementarity" between discursive formations (AK 161), it is inviting to ask what kind of accessibility relations Foucault's similarity criteria specify.[15]

First, the accessibility relation between discursive formations must certainly be reflexive since every discursive formation must be similar to itself, and thus accessible to itself. Yet the accessibility relation is neither symmetric nor transitive.[16] To see why it cannot be transitive, note that a discursive formation, df_1, can be similar to another discursive formation, df_2, with respect to one set of regularities, R_1, and similar to another discursive formation, df_3, with respect to another set of regularities, R_2, without df_2 and df_3 sharing any common, or similar, regularities. Foucault himself mentions that a discursive formation can "enter simultaneously into several fields of relations", viz., several "regions of interpositivity" (AK 159). Clearly, for these "several fields" to remain distinct, for them not to collapse into one larger region of interpositivity, it must be possible for a discursive formation to be accessible from, and have access to, at least two other discursive formations that are not accessible to one another. Finally, the question whether the accessibility relation is symmetric must also be answered negatively. This is important since it suggests that similarity between discursive formations is only a necessary but not a sufficient condition for accessibility. Certainly, similarity is a symmetric relation: if A is similar to B, then B is also similar to A. Yet when one discursive formation is subordinated to another – a possible situation mentioned by Foucault himself (AK 161) – it seems natural to allow for the possibility that the communication channel is used only in one direction. In such a case the accessibility relation is not symmetric, however similar the two discursive formations happen to be.

Before leaving the problem of communication between different discursive formations, one obvious critical question has to be addressed, at least briefly. This question turns on the apparent contradiction between Foucault's allowing for interdiscursive communication, on the one hand, and his rejection of transdiscursive statement identity, on the other hand. How can one discursive formation influence another, if no statements can cross their border? Two answers can be suggested to this question. First, one might suggest that influence in this case does not have to involve any exchange of statements. The influence of one discursive formation, say df_1, on another, say df_2, might simply involve df_2 adopting similar structures of institutions and theories. Second, recall that Foucault allows for "linguistic analogy" between statements of different discursive formations, i.e., identity "from the point of view of grammar (vocabulary, syntax, and the language (*langue*) in general)". He also speaks of this linguistic analogy as "translatability" (AK 145). Perhaps we should say that even though a statement ($é_i \in df_i$) cannot be transferred to another discursive formation, it can at least be translated into a statement of another discursive formation ($é_j \in df_j$), where the second statement ($é_j$) may then be regarded as a counterpart of the first ($é_i$).

6. THE ARCHAEOLOGICAL MODEL II: BEYOND CONTINUITY AND DISCONTINUITY

PHILOSOPHERS' *VS.* HISTORIANS' HISTORY OF SCIENCE

Needless to say, the touchstone of respectability for any conceptualization of the history of science is its ability to capture phenomena of change. That is to say, the relative success of any proposal for how to write the history of science cannot but be measured in terms of its skill in distinguishing between phases of rapid and slow change, its ingenuity in dealing with continuity and discontinuity, and its ability to separate apparent or superficial changes and tendencies from deep-going and far-reaching ones. Therefore, only by reconstructing and discussing Foucault's suggestions with respect to these topics can we put ourselves in a position where we can evaluate the strengths and weaknesses of the whole archaeological edifice.

Reviewing the literature on changes and revolutions in the history of science, published outside of France, one inevitably notices the peculiar fact that philosophical and historical scholarship have lived, by and large, separate lives. On the one hand, philosophers of science have developed a number of theories on the logic of scientific change, focussing centrally on the question of how scientific revolutions are to be accounted for, or whether or not they exist at all. Historians of science, on the other hand, have actually *done*, that is, *written* the history of science, without, however, showing much awareness of, or at least without using systematically, and testing critically, the philosophers' ideas.[17]

The negligence of historians in this area has often been deplored by philosophers, who have argued that the dearth of theory of historians predisposes them to write the history of science in terms of unreflected and philosophically long-outdated and obsolete terms.[18] Even when historians have themselves come up with some theorizing, their suggestions tend to be somewhat unsophisticated. To pick just one instance, let us take a brief, critical look at I. Bernard Cohen's monumental work *Revolution in Science* (Cohen 1985). Cohen's study is an exceptional achievement for a historian of science not only for the vastness of material covered: the study analyzes more than twenty smaller and larger scientific revolutions since the 17th century and reports an endless number of views on what constitutes a scientific revolution. But Cohen's book is also remarkable in that it amounts to one of the rare cases where a practicing historian of science seeks to spell out explicitly his views on what constitutes a scientific revolution and how it is to be identified.

The four criteria that Cohen comes up with are quickly stated. First, a revolutionary advance in science is always claimed to be such by the scientist who causes it. Second, the new idea immediately floods into the literature of the field in question. Third, "competent" historians today confirm that the new idea – or the bundle of new ideas – indeed amounted to a revolution. And fourth, scientists practicing in the same field today agree that the idea in question really meant a revolution (*ibid.*, 40–47).

As is easy to show, these criteria are highly unsatisfactory even if – in fairness to Cohen – we take note of his disarming admissions that these criteria are "somewhat subjective" and "do not cover every possible contingency" (*ibid.*, 47). In fact, it is even somewhat amusing to note that Cohen's own case-studies of various scientific revolutions in the very same book provide ample material to challenge his methodological suggestions. As concerns the first condition, one can object that, as Cohen himself impressively demonstrates, the whole conception of science as advancing in terms of revolutions and ruptures is not older than the 17th century. Obviously, the criterion is also based on the somewhat questionable views that a scientist has to have a full understanding of her work's importance, and that modesty is a negligible factor among scientists. With respect to the second criterion, the claim that the new idea must immediately flood into the literature, it has to be said that Cohen fails to tell us just how many years may come to pass for the flooding still to qualify as an immediate one. Mendel's discovery of the founding laws of genetics had to wait for almost fifty years to flood into the literature, and even Freud's ideas – and Cohen speaks explicitly of "the Freudian revolution" – were widely discussed only after a delay of some fifteen years. In the first six years after the publication of *Die Traumdeutung*, only 351 copies of the book were sold (Freud 1982: 19). Hardly surprising, Cohen's third condition, the judgment of "competent historians", fares even worse. Clearly, if there ever were a need for a good example of a circular criterion, Cohen's suggestion will do as good as any. Since the competence of a historian depends in great part on his ingenuity in identifying breaks and changes, we cannot identify revolutions by relying on the historian's alleged competence. Finally, as concerns the fourth criterion, the view of scientists today, we need to do little more than recall the well-known fact that natural scientists usually have a rather rudimentary understanding of the history of their discipline, an understanding largely based on superficial, and not markedly "competent" textbooks.[19]

Now whereas the unhealthy split between the philosophy and the history of science is fairly obvious in the academic world outside of France, such an abyss does not seem to exist in the French context. After all, in France, from Koyré to Canguilhem and Foucault, the history of science has been written by men trained in philosophy. This fact alone might be regarded as a *prima facie* reason for philosophers to acquaint themselves with French thought. Unfortunately, however, philosophers of science who have dared to delve into French thought have barely scratched the surface. This is especially noticeable in Foucault's case, who has typically been read - at least by philosophers of science - as something of an exaggerated, if not to say perverted, version of Kuhn. For instance, Larry Laudan, in his *Science and Values* (1984) refers to Foucault as a historian of science who, like Kuhn, conceives of scientific change as "*simultaneous* rather than *sequential*" (*ibid.*, 69). Since Kuhn is wrong, Foucault's excessive Kuhnianism can then only be "unintelligible ... nonsense", as Laudan informs his reader in his earlier *Progress and Its Problems* (Laudan 1977). Laudan tells us that for Foucault, "the archaeology of ideas" should be written in terms of "ruptures of human consciousness" and by "invocation of the *Zeitgeist*". Laudan concludes, unsurprisingly, that "with its twin emphases on the mystery and the opacity of human thought, with its stress on 'history as poetry', Foucaultian structuralism must rank as one of the most obscurantist historiographical fashions of the twentieth century" (1977: 241). Given such an assessment, it is hardly surprising that philosophers of science, with the notable exception of Ian Hacking, have displayed little interest in Foucault's work.

Back in France, the situation has been different. There philosopher-historians like Canguilhem and Dominique Lecourt have commented favorably on *The Archaeology of Knowledge*. For instance, Lecourt hails

Foucault's *discours de la méthode* as a major advance over Bachelard and Althusser. Lecourt commends Foucault especially for having outlined a way of writing the history of science "which refuses both the *continuity* of the subject and the structural *discontinuity* of 'ruptures'" (1975: 189), and which looks upon science and ideology "as a material, historically determinate instance in a complex social whole, itself historically determinate" (*ibid.*, 208). Canguilhem, on the other hand, stresses the importance of Foucault's conception of ideology for his own work and professes to have perhaps overlooked earlier the Foucauldian distinction between different thresholds (1988: x).

In my own reconstruction of the Foucauldian conceptualization of change in *The Archaeology of Knowledge* I shall take my lead from Canguilhem and Lecourt rather than from Laudan. In brief, I shall try to show that assimilating Foucault to Kuhn is mistaken, first, because Foucault rejects the continuity *vs.* discontinuity opposition, second because Foucault's discursive formations are quite different from Kuhn's paradigms, and third because *The Archaeology of Knowledge* has something of substance to say on the social conditions of science.

The Debate over Continuity and Discontinuity Reconstructed

In order to situate Foucault adequately with respect to the debate on scientific revolutions or discontinuities in general, and Kuhn's *The Structure of Scientific Revolutions* in particular, it seems imperative to provide a brief reconstruction of the main stages of this dispute both inside and outside of France. Developing further a recent suggestion by Peter Galison (1988), I submit that this debate can be structured around four models for the philosophy and periodization of science: the *positivist-empiristic* model, the *antipositivist* model, the *revised antipositivist* model, and the *critical postmodern* model. By and large, these four conceptions can be put in an order of succession at least as far as their becoming an object of scholarly debate is concerned.

Galison speaks of these models as constituting different types of philosophies of science. Each type is characterized by *one* "central metaphor" which, by the same token, "has deep consequences for the structure of narrative offered by the *history* of science" (*ibid.*, 198).

While the suggestion of an interrelation between different philosophies of science and different ways of writing the history of science is not new, of course, Galison manages to capture this interrelation in terms of somewhat oversimplified yet highly suggestive archaeological section representations.

The positivist-empirist model

The positivist-empirist model operates mainly with two strata, a lower and independent-determining stratum of observations and experiments and a higher and dependent-determined stratum of theories (Figure 5).

	Positivist periodization				
...	theory$_1$	theory$_2$	theory$_3$	theory$_4$	...
...		observation, experiment			...
	time ►►►				

Figure 5

The empirist or the positivist conceive of science as being founded upon accumulating observations that, as time goes by, get more precise and enlarge in scope. These observations and the observational terms in which they are formulated are taken to be independent from theory, or theoretical terms. In other words, observations provide the foundations that have to be "saved" by different, successive theories.

As Galison is right to propose, this picture of the nature of science leads to a definite view on how the history of science should be written. Since observations and experiments are the crucial stratum, and since on this level no radical breaks or reorganizations occur, the history of science is a history of accumulation and continuity. To be sure, discontinuity is met with on the level of theories, but since theories are just codifications of observations, these discontinuities are but phenomena of the surface or superstructure with no bearing upon the underlying deep continuity (*ibid.*, 202–3).

As other writers have pointed out, this picture of science also has the further historiographical consequences of encouraging a belief in progress, a search for predecessors and anticipations, a belief in the unity and progressiveness of science, and the writing of the history of science as a history of methods and results (Agassi 1963: 33–40; Fichant 1969: 96–97). First and trivially, a conception of science working with the postulate of accumulation will have no problems defending the progressiveness of science. Second, because of its continuity assumption, this model suggests looking for predecessors. Furthermore, since the continuity assumption derives from the belief in an independent stratum of observations, it is also tempting to think that – at least in principle – many later-day observations could actually have been made, or were actually made, much earlier. Third, given these premisses it is also natural to speak of the unity of science as grounded on the common observational, theory-independent and discipline-neutral stock of observations. And fourth, the positivist section metaphor encourages the historian to concentrate on methods for, and the results of, doing science: science is a continuous piecemeal revision of theories, a piecemeal revision into which major political and social events can intervene only as temporary obstacles.

Whereas the French *locus classicus* of this kind of historiography and philosophy is the work of Duhem, in the English speaking world, the underlying philosophy is of course foremost represented by logical empiricism which dominated philosophy in the United States until the 1960s. While little historical work has actually been done by members of this line of persuasion, it has been suggested that George Sarton's work on the history of science displays in practice a naive form of what logical empirism or earlier and later positivism formulate as a much more sophisticated theory (Fichant 1969: 67–70).

The antipositivist model

The interesting difference between the English speaking world and the French scene is that the positivist metaphor or section drew fire much earlier on the old continent. Koyré's study of Galileo came from the press in 1939, and between 1928 and 1940, Bachelard published altogether ten books outlining and defending the antipositivist philosophy of science as well as its historiographical counterpart.

Central in the antipositivist model is the reversal of the order between theory and observation; now theories form the fundamental and determining stratum, and observations lose their theory-independence (Figure 6).

Antipositivist periodization

...	observation_1	observation_2	observation_3	...
...	theory_1	theory_2	theory_3	...

time ►►►

Figure 6

To be sure, to make Bachelard a genuine proponent of this picture is something of an oversimplification; his theory of the relations between theoretical and experimental physics allows for more than unidirectional influence between the two strata. Nevertheless, placing him under the antipositivist *Idealtyp* can be justified by the fact that the way he was read in the France of the fifties and sixties fits well with this figure.

As few will need to be reminded, the antipositivist "revolution" in the English speaking world occurred in the early sixties when Kuhn's *Structures* appeared. Kuhn, Feyerabend and their fellow travellers argued that observation is inevitably "theory-laden", and that a theory-neutral observational "protocol language" à la Carnap is an impossibility. It follows of course that the history of science is the history of (bundles of) theories or more general theoretical and methodological assumptions. Furthermore, now that observations are refused the role of the underlying stratum common to successive paradigms or grand theories, the antipositivists tend to go further - or at least are interpreted by their critics as tending to go further - and deny that there is, or even could be, any other substratum that connects successive sets of general assumptions or paradigms. The endpoint of this argument thus is that adherers to different paradigms live in "different worlds", and that the question of whether science is progressive or not is a futile one (Galison 1988: 205).

As to the historiography written according to this model, I have already noted that for instance the Kuhnian model has inspired few if any historians in their own research. Or, if it has inspired them, at least few if any actually employ the Kuhnian scheme of 'paradigm - crisis - new paradigm'. In fact, in his own substantial historical work on black-body radiation, not even Kuhn himself draws on this model. As far as other antipositivist studies of the history of science are concerned, the major contributions come again from France, to wit, the works of Koyré (who inspired Kuhn), Bachelard and Canguilhem.[20] These authors, however, are not bound to any one scheme for writing the history of science, and their antipositivism surfaces more in the forms of a general attention to problem situations of a science of a given period, a rejection of the search for predecessors and anticipations, a heightened awareness of the influence of metaphysical views upon science, and a general willingness to speak of discontinuities and breaks.

Adding strata: the antipositivist revisionism

To proceed to the next stage of the debate (a stage not distinguished by Galison), a first wave of nonpositivist critics of the antipositivist model used mainly two strategies, often in tandem; they introduced new strata into the antipositivist section drawing and they argued for more fuzzy borders between successive paradigms (Figure 7). In combination, these strategies were meant to preserve the rationality and progressiveness of science. As critics from Popper, Lakatos and Toulmin, to Laudan, Canguilhem and Hacking saw it, the Kuhn-Feyerabend model of revolutionary scientific change reduces the latter to being "a mystical conversion which is not and cannot be governed by rules of reason ... a kind of religious change", and "a matter for mob psychology" (Lakatos 1974: 93, 178).[21]

Revised antipositivist periodization

. . .	observation$_1$	observation$_2$	observation$_3$	. . .
. . .	theory$_1$	theory$_2$	theory$_3$	. . .
. . . . . .	disciplinary principles / empirical problems styles of reasoning / research mentalities			. . . . . .

time ►►►

Figure 7

Since some of these revisions of the antipositivist section drawing will turn out to be helpful in my reconstruction of Foucault, let us take a closer look at Stephen Toulmin's, Larry Laudan's and Ian Hacking's new proposals.

Toulmin's *Human Understanding* employs both of the two main revisionist strategies I have distinguished. As concerns the argument from fuzziness, he argues against Kuhn that the changeover from Newtonian to Einsteinian physics cannot be described as "a complete rational discontinuity" (Toulmin 1972: 103). Einstein's work did not constitute "a fullscale scientific revolution" because every single one of the modifications that Einstein suggested was discussed in detail, and accepted only after an intensive debate. Indeed, the more radical the proposed revisions are that a scientist puts forward, the more prolonged will the ensuing debate be (*ibid.*, 104). Furthermore, Toulmin plays out "the writings of T.S. Kuhn the historian" against "the theories of T.S. Kuhn the philosophical sociologist". As Kuhn's own work on Copernicus makes abundantly clear, Toulmin alleges, "... the so-called "Copernican Revolution took a century and a half to complete, and was argued out every step of the way". Even though it resulted in something of a radical change in "physical and astronomical *ideas and theories*", the debate over these ideas "implied no comparable break in the intellectual *methods* of physics and astronomy" (*ibid.*, 105). Thus we do best to follow political theorists, and abolish the use of the notion of revolution, at least for explanatory purposes, altogether. The distinction between normal and revolutionary change turns out "to have little real theoretical significance": "At most, the two sorts of conceptual changes differ only in degree; and certainly they must, in the last resort, be accounted for in terms of the same set of factors and considerations" (*ibid.*, 118).

The new substratum that Toulmin wishes to introduce, and the substratum that is meant to negotiate between different theories and ideas, are "disciplinary principles". Other than "theoretical principles" like Newton's Principle of Universal Gravitation or Mendel's Principles of Segregation and Recombination, disciplinary principles define "the basic intellectual goals of a science", e.g. like the explanation of physiological processes in terms of chemical ones. Toulmin suggests that we confine Kuhnian paradigms to the level of theoretical principles, and concludes on this basis that the incompatibility between Newtonian and Einsteinian physics did not extend to the deeper level of disciplinary principles: "... supporters of the two positions shared enough disciplinary aims for them to be able to discuss, in a vocabulary intelligible to both sides, which of the two theories 'did the better explanatory job' for theoretical physics" (*ibid.*, 124). Radical incompatibility, like a disagreement over disciplinary principles, does occur, e.g. Goethe's disagreement with Newton over colors, but such cases are not really diachronic breaks within a discipline. Precisely because Goethe does not subscribe to Newton's disciplinary principles, he does not belong to the same discipline of theoretical physics (*ibid.*, 124–25).

In Laudan's *Progress and Its Problems* (1977), the new substratum consists of "empirical problems" rather than disciplinary principles. Laudan believes that the historian of science can identify a slowly changing set of empirical problems that form "a *permanent* fixture of the science changes" (*ibid.*, 140). While "research traditions" like Aristotelianism, Darwinism or Freudian psychology change or are displaced and replaced, Laudan submits, this permanent fixture remains and provides the touchstone with respect to which the relative success of different research traditions can be evaluated. Different research traditions can be compared, since "with respect to any two research traditions (or theories) in any field of science, there are some joint problems which can be formulated so as to presuppose nothing which is syntactically dependent upon the specific research traditions being compared" (*ibid.*, 144). Laudan believes that the introduction of this new substratum enables him to provide a partial justification to both "revolutionary" and "gradualist" conceptions of the development of science: discontinuities occur on the level of "explanation or problem solution", whereas on the level of empirical problems the assumption of continuity or slow accumulation is justified (*ibid.*, 140).

However, this partial rehabilitation of the discontinuist viewpoint is to be taken with a grain of salt, since Laudan also employs the second strategy of the revisionist antipositivist. Laudan rejects the Kuhnian *Gestalt* switch not only because of his assumption of paradigm neutral empirical problems; another line of argument of his weakens the assumption of discontinuity on the theoretical level itself. This line has two ingredients: on the one hand, Laudan allows for extensive change already within a research tradition; on the other hand, he suggests the co-existence of competing research traditions. First, *Progress and Its Problems* modifies Lakatosian research programmes by "relativizing the 'essence' of a research tradition with respect to time" (*ibid.*, 100). In other words, the hard core of the most central, general assumptions of a research tradition is not sacrosanct, but can change over time:

> ... when it can be shown that certain elements, previously regarded as essential to the whole enterprise, can be jettisoned without compromising the problem-solving success of the tradition itself, these elements cease to be a part of the "unrejectable core" of the research tradition (*ibid.*, 100).

Second, Laudan rejects the Kuhnian notion that a new paradigm emerges only after a prolonged crisis which lead to the abandoning of its predecessor. Instead, Laudan believes in a "perennial co-existence of conflicting traditions", claiming that it is just this coexistence which makes "the focus on revolutionary epochs so misleading" (*ibid.*, 136). While not denying the applicability of the concept of revolution to certain changes in the history of science, Laudan sees revolutions as phases where a large number of scientists of a given field start to take seriously a rival research tradition because of its success in problem-solving (*ibid.*, 138).

The third revisionist notion, or new substratum, is "style". To be sure, only some uses of this concept can be regarded as revisionist, since the same term has also played the role of an important metaphor in the writings of some antipositivists themselves.[22] More clearly as a critical weapon against the Kuhnian conception, the notion of style is employed by Alistair Crombie and Ian Hacking.[23]

Crombie wants to apply to the study of the history of science a distinction familiar from *Annalist* historiography, to wit, the distinction between the history of ideas or theories and the history of (research) mentalities (Crombie 1981, 1986, 1988). The later takes as its subject matter very general commitments concerning conceptions of nature, science and morals, and relates these to one another via a "taxonomy of scientific styles":

> ... [1] the simple *postulation* established in the mathematical sciences, [2] the *experimental exploration and measurement* of more complex observational relations, [3] the *hypothetical construction of analogical models*, [4] the ordering of variety by *comparison and taxonomy*, [5] the *statistical analysis* of the regularities of populations and the calculus of probabilities, and [6] the *historical derivation* of genetic development. The first three of these methods concern essentially the science of individual regularities, and the second three the science of the regularities of populations ordered in space and time (Crombie 1981: 284).

In a number of papers, Hacking (1982, 1983a, 1985; cf. 1983b: 55, 71) has spelled out the philosophical implications of Crombie's notion of "styles of reasoning". He uses this conception for two purposes: in order to defend some sort of relativism and in order to preserve more continuity than the original antipositivist picture of science seems to allow for. Hacking proposes that the assumption of styles of reasoning forces us to allow for relativism, simply because "the very candidates for truth and falsehood have no existence independent of the styles of reasoning that settle what it is to be true or false in their domain" (1985: 146). However, the forms of knowledge which different styles of reasoning define are not to be equated with Kuhnian paradigms. Whereas Kuhnian paradigms can be shared by only a small group of researchers, styles of reasoning are "vastly more widespread" (*ibid.*, 149). They evolve much more slowly, can combine and overlap, separate, and – most important of all – accumulate:

> I can agree with [Kuhn:] ... *Knowledge* is less cumulative than we thought. But other things accumulate. In addition to technology and experimental technique, we can add one style of reasoning to another ... and if a style dies out, the death is slow and often passes by unnoticed (*ibid.*, 148).

For Hacking as for Crombie, styles of reasoning thus provide an additional substratum, below theories, or paradigms, a substratum whose breaks and reshuffles need not, and typically do not, coincide with changes on these upper levels.

Laudan and Foucault as postmodernists

Hacking's as well as Peter Galison's work (Hacking 1983; Galison 1987) is also the natural starting point for approaching, finally, the critical postmodern model, as Galison calls it. Even though this model too emerged from a critical reflection upon antipositivism, it deviates more radically from its basic assumptions than the revisionist viewpoint. *Eo ipso*, the criticism of the postmodernists thus also applies to the revisionists.

By and large, the postmodern approach is characterized by two steps: it denies that (revolutionary) changes in science affect all strata simultaneously, and that certain elements, like observations or theories, are always situated on the determining, deeper level. In fact, Galison argues that the antipositivist model is but a "flip-side version" of the positivist section, in that both models "have a well-established hierarchy that lends unity to the process of scientific work ... in their mirror reflections there is a good deal of similarity" (1988: 207). For Galison there is no "reductive hierarchy" to be established between different levels of physics, i.e. "theoretical traditions", and "experimental traditions", and therefore "it is quite possible for continuity to persist at one level while being broken at another" (*ibid.*, 209). (Figure 8)

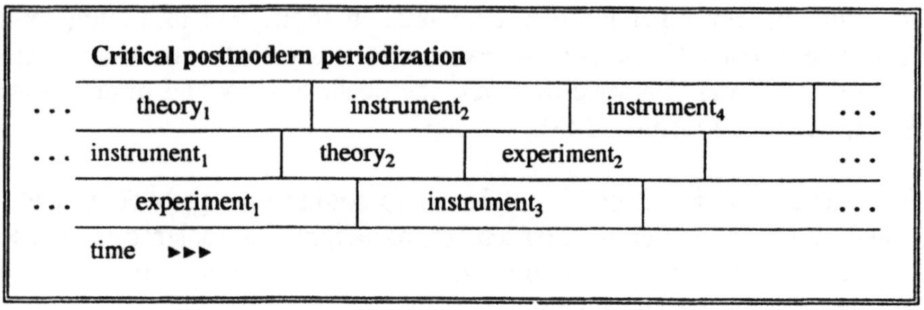

Figure 8

Contrary to what Galison implicitly alleges, this critical postmodern model does not only emerge with "the new experimentalism", i.e. with philosophers of science like Hacking and Galison himself who redirect their attention from the way theories *represent* reality towards the problems connected with how scientists *intervene* in it by way of experiments.[24] This model also surfaces in Laudan's *Science and Values* (1984) and, much earlier – alas – in Foucault's *Archaeology*.

Laudan is not so much concerned with the interplay of theory, experiment and instrument, as with the interrelations of the three main components of a Kuhnian paradigm, i.e. theories, methods, and aims. As Laudan sees it, philosophers of science tend to make two mistaken assumptions concerning these components. First, and here Laudan includes pre-Kuhnian as well as Kuhnian conceptions, philosophers suppose that the interrelation between the three components is a hierarchical one, such that scientific debates over theories are resolved at the level of methodology, and methodological debates at the level of values. Naturally, debates over values must thus inevitably remain undecidable on rational grounds (1984: 23–43). Second, Laudan claims that Kuhn and like-minded thinkers, including Foucault, fall prey to "the covariance fallacy", that is, the fallacy of assuming that all three components can only be replaced simultaneously, not successively (*ibid.*, 43–50). In other words, the assumption that theories, methods and aims come in one "inseparable package", making paradigm changes turn out to be "abrupt and global ruptures", is wrong (*ibid.*, 70–71).

Laudan's arguments against Kuhn are directed against precisely this inseparable package view. Drawing on examples from the history of physics and chemistry - examples earlier interpreted by Kuhn as constituting paradigm shifts - Laudan (*ibid.*, 76–77) shows that the changes involved are not instantaneous ones from

(time$_1$:) theories$_1$ - methods$_1$ - aims$_1$

to

(time$_2$:) theories$_2$ - methods$_2$ - aims$_2$

but rather step-by-step revisions of the three constituents. Laudan suggests that only a superficial historian - Laudan baptizes him "Tom" - can see abrupt change where in fact the change is almost always gradual. Tom will be so struck by the shift from time$_1$ to time$_2$, perhaps happening in just one decade, that he will be blind to the intermediate stages:

> If Tom decides to call the view that scientists eventually come to hold 'Paradigm 2', and the view from which they began 'Paradigm 1', then he will be able to document the existence of a massive paradigm shift between what (at our remoteness in time) appear to be conceptually distant and virtually incommensurable paradigms (*ibid.*, 78).

To correct these mistakes, Laudan proposes his own "reticulated model of scientific rationality", which allows not only for different ways in which theories, methods and aims constrain each other (aims must harmonize with theories, and they justify methods, which in turn exhibit the realizability of aims, and justify and are constrained by theories), but also for slow erosions and sudden overturnings. (Figure 9)

With this rough outline of the postmodern models of Galison and Laudan, we can turn to the question whether Tom would feel comfortable in the company of Michel. Are we to conceive of a Foucauldian "rupture" or "discontinuity" as a Kuhnian "irreversible Gestalt-shift", or "conversion experience", as an instantaneous break that renders discursive formations incommensurable, as a barrier that makes researchers working in historically subsequent discursive formations live in different worlds? Some of Foucault's statements indeed suggest this kind of

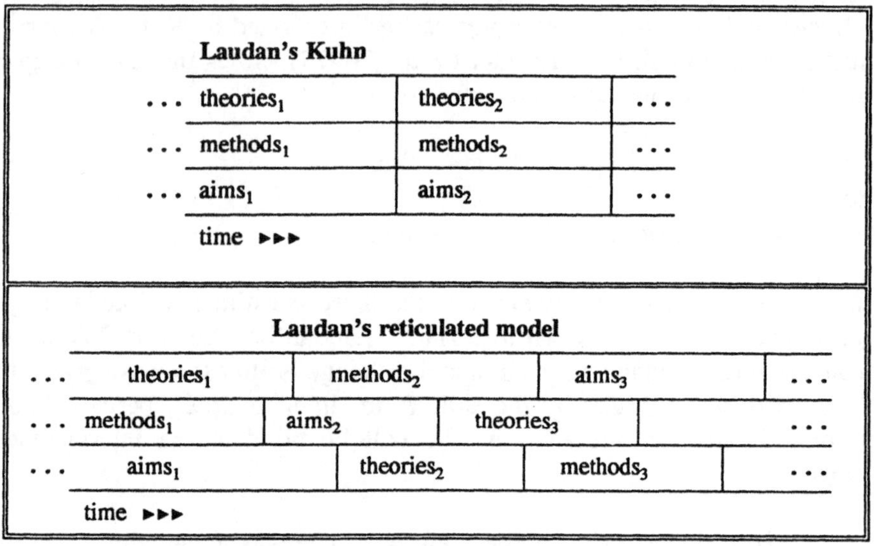

Figure 9

interpretation. For instance, *The Order of Things* speaks of "the suddenness and thoroughness with which sciences were sometimes reorganized", and the same book tells us that after the "great discontinuity" at the beginning of the nineteenth century, "Classical thought ceased ... to be directly accessible to us" (OT xii, xxii, 304). As I have already indicated, my answers to the questions posed are nevertheless negative. They must be negative not only because of Foucault's explicitly stated theory of rupture in *The Archaeology of Knowledge*, but they must also be negative in the light of his historical-archeological investigations, i.e., *Madness and Civilization*, *The Birth of the Clinic*, and *The Order of Things*.

To review briefly the latter works, what *Madness and Civilization* studies is not the immediate, break-like emergence of our modern conception of madness, but rather the gradual, step-by-step emergence of this conception from the Renaissance to the 19th century. In *The Birth of the Clinic*, Foucault studies the rupture in medical thought roughly between 1780 and 1825. Again the change under scrutiny is gradual, not immediate: the period under investigation is divided into four phases and the transformations taking place from one phase to the next are everything but revolutionary. It is only when taken together that these phases illuminate the radical, rupture-like change in medical thinking between the outer limits of the whole period. Finally, *The Order of Things*, even though it speaks of "two great discontinuities in the *episteme* of Western culture: the first inaugurates the Classical age (roughly half-way through the seventeenth century) and the second, at the beginning of the nineteenth century" (OT xxii), suggests that the emergence of modern thought – the second rupture – is "possible to follow step by step" (OT 217). Unfortunately, however, in his best known work, Foucault does not go as far in this analysis as he does in his two earlier books. Yet even here it must be obvious, even to a superficial reader, that the change from the classical to the modern period is not understood as instantaneous and package-like. After all, the transformation is dated roughly between 1775 and 1825 and a phase of gradual transition is explicitly allowed for. In this phase, the pillars of Classical thought are replaced gradually (OT 221).

The distance between Tom and Foucault increases still more radically, if we turn to Foucault's more theoretical and programmatic texts. First of all, it deserves to be mentioned that the transition from one discursive formation to another does not involve the Kuhnian scheme of 'normal science – crisis – normal science'. This is because for Foucault "the opposition between periods of stability or of universal convergence and moments of effervescence when minds enter into crisis, ... when all notions are revised, overturned, revivified, or for an indefinite time, fall into disuse" is the most important, and worst of the "ill-considered oppositions" that haunts traditional history (1978: 17). Kuhn's cyclical picture of the development does not fit the archaeological framework, since for the latter it is axiomatic that the "appearance and disappearance of positivities, the play of substitutions to which they give rise, do not constitute a homogeneous process that takes place everywhere in the same way" (AK 175). It must also be mentioned that Foucault never denies that researchers of successive discursive formations are – in principle – able to understand one another. What Foucault does claim is that those discursive formations that are separated from us by a rupture are not "directly [!] accessible to us" (OT 304). This statement must, in the light of Foucault's *Archeology* as well as in the light of his historical studies, be read as a warning against anachronistic assimilations of earlier scientific concerns to our own, rather than as a plea for incommensurability.

Furthermore, in the *Archeology*, we read that a rupture "is not an undifferentiated interval - even a momentary one - between two manifest phases; it is not a kind of lapsus without duration that separates two periods". Rather, a rupture between two successive discursive formations is "always a discontinuity specified by a number of distinct transformations" (AK 175). Foucault also writes that "rupture is the name given to transformations that bear on the general rules of one or several discursive formations" (AK 177), or that a "discontinuity ... is a play of specific transformations different from one another (each one having its conditions, its rules, its level) and linked among themselves according to schemes of dependence" (1978: 13).

To go into greater detail, we can start from the observation that archaeology distinguishes between different forms of change according to the levels or layers on which they are situated. Change can be studied on the level of statements, on the level of the four series (objects, modalities, concepts, strategies), on the level of their regularities, and on the level of interrelations between different contemporaneous discursive formations (AK 171).

For instance, on the level of strategies, or "theoretical options", Foucault distinguishes between seven different types of transformations within a discursive formation: A theory T_2 can derive from, or be motivated by, a theory T_1 by "deduction or implication", "generalization", "limitation" of T_1, "passing to the complementary" (e.g. theories of formal languages motivating the study of natural languages), "passing to the other term of an alternative", (e.g. from action by contact to action at a distance), "permutation of dependencies" (e.g. grounding mathematics in logic leads over into the attempt to ground logic in mathematics), and "exclusion or inclusion" (1978: 11).

Furthermore, Foucault tells us that to describe the change from one discursive formation to another must include the identification of transformations on several levels: "[1] how the different elements of a system of formation were transformed (...); [2] how the characteristic relations of a system of formation were transformed (...); [3] how the relations between different rules of formation were transformed (...); [4] lastly, how the relations between various positivities were transformed (...)" (AK 172). Using the example of object formation, and employing self-explanatory abbreviations (with 'conc' for concepts and 'form' for formation), these different levels can, more succinctly, be rewritten as follows:

[1] *surface of emergence* ▸ $\textit{surface of emergence}_{new}$
[2] *Rel(surface, authority)* ▸ $\textit{Rel}_{new}\textit{(surface, authority)}$
[3] $\textit{Rel}(\textit{form}_{obj}, \textit{form}_{conc})$ ▸ $\textit{Rel}_{new}(\textit{form}_{obj}, \textit{form}_{conc})$
[4] $\textit{Rel}(\textit{df}_m, \textit{df}_n)$ ▸ $\textit{Rel}_{new}(\textit{df}_m, \textit{df}_n)$

[3] here is of course well in line with my earlier claim that for archaeology the relation between the different series is not fixed, i.e. that archaeology does not assume, *pace* Laudan, any fixed hierarchy between different strata.

Moreover, let us also recall that a discursive formation is "a scheme of the correspondence between several temporal series", a view which I interpreted to mean that for us to speak of a discursive formation it is not necessary that the time limits of all of the four series coincide. In the central chapter on "Change and Transformations" of *The Archaeology of Knowledge*, further evidence to the effect that the four central ingredients of a discursive formation do not change wholesale, that is to say, that they do not form one "inseparable package", is easily forthcoming. It is true that Foucault claims that "the appearance of a discursive formation is often correlative with a vast renewal of objects, forms of enunciation, concepts, and strategies" (AK 171), but not only does he hasten to add a counterexample ("General Grammar was established in the seventeenth century without much apparent alteration in grammatical tradition"), but he also writes that the replacement of one discursive formation by another does not entail "that all objects or concepts, all enunciations or all theoretical choices [of the earlier discursive formation] disappear" (AK 173).

It is also important to note here that archaeology regards the displacement of discursive formations as "rare ... events", and that although they are its central topic they are by no means the only one (AK 171). Indeed, archaeology concerns itself also with "continuity, return, and repetition", too:

> ... elements – or several of them – may remain identical (...), yet belong to different systems of dispersion, and be governed by distinct laws of formation. One can find in such phenomena therefore: elements that remain throughout several distinct positivities, ...; elements that are constituted, modified, organized in one discursive formation, and which, stabilized at last, figure in another; elements that appear later, as an ultimate derivation in a discursive formation, and which occupy an important place in a later formation (...); elements that reappear after a period of desuetude, oblivion, or even invalidation (...) (AK 173–75).

As all of this evidence suffices to show, Foucauldian archaeology knows of no fixed hierarchical relations between different layers, does not treat different layers as forming an "inseparable package", does not assume "incommensurability" and rejects explicitly the Kuhnian cyclical model of the history of science. Thus Foucault is as clear an instance of the critical postmodern model of science as any.

The Archaeological Dimension

It hardly needs to be argued at length that to work with fairly abstract models like Galison's types of periodizations easily leads one to overlook important differences between those philosophers of science that are grouped together as adherers to one and the same 'model'. Indeed, Galison's, Laudan's and Foucault's central interests with respect to science differ considerably, despite their common "postmodernism". Galison is mainly concerned with showing that experimentalists in the physical sciences are not mere underworkers to theoretical physicists, i.e. that experimentation has a "life of its own", and that a historical study of the instrument-building tradition of science should be related to the wider context of material culture (Galison 1988: 208). Laudan's main interest, on the other hand, is the defence of science as a rational enterprise that only in exceptional cases stands in need of sociological reconstruction. Laudan tries to achieve this goal by developing a conception of rationality according to which rationality itself can change in a rational way. Here it remains to show how Foucault's main problem differs from both Galison's and Laudan's.

That Foucault occupies himself with topics somewhat remote from the questions that philosophers of science debate can be most easily demonstrated with reference to what I propose to call 'The philosophy of science questionnaire', or *PSQ* for short. The *PSQ* demands of the proband to define her position with respect to the role, function and working of "guiding assumptions", rationality, and methodology in science, where by guiding assumptions I understand things like Thomas Kuhn's paradigms, Imre Lakatos's research programmes, Larry Laudan's research traditions, and Paul Feyerabend's global theories (Laudan *et al.* 1986: 161–62). Using the initials of these four leading figures, one can summarize their answers to the *PSQ* in the following table. (Figure 10)[25]

It is the impossibility of filling out the *PSQ* for Foucault which should have made commentators more careful in their eagerness to assimilate Foucault's work to Kuhn's historical and philosophical studies. Perhaps one can say that Foucault's texts suggest answering negatively to (*1*), (*2*) and (*7*) and positively to (*12*), but even here one feels unsure whether the guiding assumptions of Kuhn & co. can really be paralleled to the four series of a discursive formation.

Of course this negative result can lead to two different attitudes with respect to Foucault's work. Some might take the absence of clear answers to the *PSQ* simply as the final confirmation of their already entrenched view that we are indeed dealing here with "one of the most obscurantist historiographical fashions of the twentieth century" (Laudan 1977: 241). More charitable philosophers will argue that the difficulty in squeezing Foucault into the *PSQ* is simply as clear a sign as any that the *PSQ* problems are not his. These philosophers will hold that without a proper grasp of Foucault's – or anyone's – central interests, criticizing and rejecting him "is not unlike playing one of those parlour games in which one is given an answer (often a bizarre one), without knowing the question to which it is an answer!" (Laudan 1977: 176).

To turn then to what I take to be the main problem for archaeology, we must come back to the notion of threshold. I already mentioned the idea of various thresholds in an earlier chapter on Foucault's criticism of

	TK	PF	IL	LL
(1) guiding assumptions (GAs) are explicit at the outset	n	n	y	y
(2) the core of a set of GAs can change piecemeal . . .	n		n	y
(3) comparison between different sets of GAs is the rule rather than the exception	n	y	y	y
(4) new sets of GAs are seriously explored even before they succeed	n	y		y
(5) given a successor set of GAs (S*) and its predecessor (S):				
- S* accommodates S's explanatory successes	n	n	y	n
- S* is more general than S		n	y	
- S* is as precise as S	n/y			
- S* accommodates S's solved empirical problems . . .	n/y			n
- S* solves S's anomalies		n	y	n
- S* accommodates S's observational consequences . .	n		y	
(6) disputes over GAs occur constantly	n			y
(7) during a change in GAs (a scientific revolution) GAs change abruptly and totally	y			n
(8) opponents of different Ss cannot easily communicate: the meanings of their observational terms are different, they have different problems and employ similar language and experiments in different ways	y	y	n	n
(9) scientists switch to a different set of GAs because of propaganda	n	y	n	n
(10) methodological rules are formulated as a matter of routine	n			y
(11) they always change when GAs change	y/n		n	n
(12) they change over time	y		n	y

Figure 10

over-simplified oppositions in general, and Bachelard's and Althusser's strict dichotomy between science and non-science or ideology, in particular. Now we have to see in greater detail how crucial this distinction is to the whole archaeological enterprise.

The first threshold (the threshold of positive discourse), i.e. the emergence of a discursive formation out of the discursive field in general, was defined above as the arising of a more stable interplay between the four series. More can and needs to be said about the other three thresholds, and the types of discourse they characterize.

The second threshold, the threshold of epistemology, issues in what we might call an 'epistemology-based discourse'. *The Archaeology of Knowledge* tells us that at this stage within a discourse, "a group of statements is articulated, claims to validate (even unsuccessfully) norms of verification and coherence, [a group which] ... exercises a dominant function (as a model, a critique, or a verification) over knowledge" (AK 186–87). To flesh out this sketch it is inviting to relate it to Kuhn's preparadigmatic phase as well as to some other notions in the philosophy and anthropology of science.

As concerns Kuhn, it can be pointed out, first of all, that his notions of "immature" or "preparadigmatic" science do not differentiate between the different types of discourse that Foucauldian archaeology distinguishes in the "prehistory" of a science. In the first edition of the *Structure*, Kuhn, like Bachelard and Althusser, works only with a simple dichotomy, i.e. a dichotomy between fields having one (or very few) paradigms, and fields that are, in the absence of a paradigm, "relatively free" in that there is "no standard set of methods or of phenomena that every ... writer [feels] forced to employ and explain" (1970: 13). However, in the 1970 "Postscript" to the second edition of *The Structure*, Kuhn introduces a modification: the difference between maturity and immaturity is no longer linked to the presence of a paradigm:

> ... it may help to point out that the transition [from immaturity to maturity] need not (I think now should not) be associated with the first acquisition of a paradigm. ... What changes with the transition to maturity is not the presence of a paradigm but rather its nature. Only after the change is normal puzzle-solving research possible (*ibid.*, 179).

This quotation suggests that we can distinguish between three phases: first, a phase where models of research are so vague, general and conflicting that we cannot yet speak even of a paradigm; second, a phase where paradigm exemplars exist, but not yet as well-defined puzzles; and third, a phase where the number of paradigms has been drastically reduced, and the "normal puzzle-solving research" has become possible. The second stage of this conception might then be looked upon as corresponding roughly to Foucault's epistemology-based discourse. Other notions naturally situated on this level are perhaps Toulmin's disciplinary principles, Crombie's and Hacking's styles of reasoning or research mentalities.[26]

Moving on to the "threshold of science", Foucault's initial characterization of the resulting 'scientific discourse' as one which "obeys a number of formal criteria ... [and] certain laws for the construction of propositions" (AK 187), is again rather uninformative. However, to give this form of discourse more content, it helps to note that Foucault later situates Bachelard's and Canguilhem's epistemological history of science on this level (AK 190), a fact that suggests that "scientific discourse" is characterized by the presence of one (or very few) problematics. Equating problematics with Kuhnian paradigms as "disciplinary matrixes", it thus seems natural to say that on this level there exists a "community of specialists" who are committed to certain "symbolic generalizations ... which can readily be cast in a logical form like $(x)(y)(z)\varphi(x,y,z)$", "beliefs in particular models" which among other things "supply the group with preferred or permissible analogies and metaphors", cognitive values like commitment to accurate predictions, simplicity, consistency, and a fixed set of exemplars (Kuhn 1970: 182–91). It also seems natural to situate Toulmin's aforementioned theoretical principles on this level.

Finally, suffice it to mention that Foucault wants to make a further distinction by differentiating between the threshold of science and the "threshold of formalization". As the term already indicates, the formalized discourse is one in which theories are formulated in a logical or mathematical language (AK 187).

Now as it stands up to this point, the Foucauldian distinction between these thresholds and types of discourse is an interesting suggestion for a more fine-grained grid for the study of the emergence and development of mature, scientific disciplines. It cannot be stressed too much, however, that this way of reading the threshold theory is still one-sided. What this reading overlooks is that the Foucauldian distinction between the various forms of discourse is not only a distinction between successive stages of the prehistory and history of a scientific discipline, but also - and first and foremost - a distinction between different perspectives with respect to, or different strata of, one (or several) scientific discipline(s).

To begin with, the *Archaeology* claims that the chronology of the different thresholds and types of discourse "is neither regular nor homogeneous" (AK 187). This is to say, first, that not all fields of learning pass through these different stages simultaneously, "thus dividing up the history of knowledges into different stages". Second, the order in which different discourses pass through these stages does not follow a fixed scheme: different discourses cross different thresholds at different intervals, and for some discourses the passing of two or more of these thresholds can coincide. For example, mathematics has crossed all of the four thresholds at once (AK 188).

Third, and most importantly, even though for each discourse one can identify the moments when it crosses one of these thresholds for the first time, the structure it obtains at these thresholds is not necessarily abolished when another, later or higher, of the thresholds is passed. Furthermore, we must reckon not only with thresholds where a discourse receives orders (of the four kinds) for the first time, but also with thresholds where for one and the same level, the orders are changed or replaced. Put differently, the specific structure that a discourse receives at a given level is preserved even when the following 'higher' level emerges. Subsequently, what originally came about - at least typically - successively, now interacts, and constitutes different interrelated strata that can, but need not necessarily, develop simultaneously. (Figure 11)

In other words, changes on any of the four levels can, but need not necessarily, go hand in hand. Thus the introduction of, or the change in, formal mathematical theories need not be accompanied by a change of paradigms, as evidenced by Maxwell's mathematization of Faraday's field theory. The introduction of, or the change in, paradigms need not coincide with changes in disciplinary principles, as Toulmin shows concerning the debate between Newtonians and Einsteinians. However, the introduction, or emergence, of a new style of reasoning can be

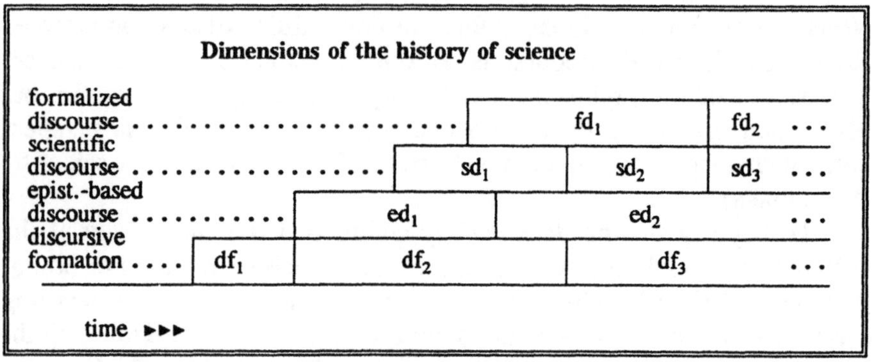

Figure 11

accompanied by the first introduction of a paradigm and a mathematical tool, as Ian Hacking's study of the emergence of probability in the 1660s demonstrates (Hacking 1975). Finally, the structure of the epistemology-based discourse can change without paradigms or formalisms being replaced (immediately); for instance, the image of (physical) knowledge of some physicists has changed as a result of the atom bomb, or of the Star Wars program, without this change having resulted in any new paradigm or formalism.

Having already seen how archaeology defines the relation of the level of discursive formations to weakly structured discourses, the question that remains is how the introduction and change of a discursive formation relates to other changes. To answer this question, we have to recall that a discursive formation is not just a series of series of objects, enunciative modalities, concepts and themes or strategies, but also the interrelation of regularities of *social* conditions that constitute these series. (This is obvious in *Madness and Civilization*, *The Birth of the Clinic*, and *The Archaeology of Knowledge*, but obscure in *The Order of Things*.) Thus we saw that central in the constitution of objects like madness were, e.g., the role and function of institutions like the family and the working place, as well as the authority of the doctor, the judge and the priest; crucial for the emergence of a specific series of enunciative modalities in medical discourse were the social structure within science and the role of the scientist in society; important for the role of certain concepts like the 'spring of air' are specific orderings of statements which in turn reflect the agonistic struggle between different schools and different disciplines (Boyle *vs.* Hobbes); and, finally, all important for the ending of debate over theory choice are competition for important alliances and the intervention of non-scientific interests. What all this suggests, therefore, is that the level of discursive formations is the stratum of the social conditions for the emergence of a structured discourse within which – perhaps only at a much later date – emerge one or several disciplines with varying disciplinary principles, paradigms and theories.

Obviously, given this role of discursive formations, it is then only natural to assume that with the emergence of further structures like paradigms, the earlier level of discursive formations has by no means become superfluous. As these new structures evolve, there are new regularities and series to be accounted for, series that do not neatly coincide with changes in the social conditions of series of objects or theories. Indeed, Foucault claims that these new series do not necessarily have the same limits as successive systems of social conditions, i.e. discursive formations (AK 188). It is clear, e.g., that the existence span of a discursive formation is considerably longer than the span of theories of even paradigms; after all, a discursive formation is defined as a series of even conflicting theories, objects, modalities and concepts. On the other hand, it is also compelling to allow that there are cases where paradigms or theories remain, regardless of a change in the social conditions; mathematics or theoretical physics are as good examples as any to establish this point.

7. ARCHAEOLOGY OF KNOWLEDGE AND OTHER HISTORIES OF SCIENCE

Discursive formations as series of series of objects, enunciative modalities, concepts, as well as their conditions, are thus themselves just one series, one level to be described by histories of science. Not only is this claim strongly suggested by my interpretation above, but it is also stated straightforwardly by Foucault himself. In fact, he distinguishes between three types of histories of science, according to the level or series of events they are primarily concerned with. Thus the history of mathematics writes the history of the series of formal theories, where typically later theories are generalizations of earlier ones (AK 189). Foucault does not mention examples of such histories, but one might mention the work of Jean Cavaillès as an outstanding example of this type of historiography. A second way of writing the history of science is the study of the emergence of problematics or paradigms. As Foucault sees it, "G. Bachelard and G. Canguilhem have provided models of this kind of history" (AK 190). Interestingly enough, *The Archaeology of Knowledge* claims that this kind of history can only be written by taking a normative perspective, that is by judging the past of science in terms of its present ("this description takes as its norm the fully constituted science" (AK 190)). Finally, the description of the emergence of a discursive formation as well as its acquiring of epistemological principles, Foucault reserves for the archaeology of knowledge (AK 191).

Here is seems inviting to go further and suggest that the distinction between the four levels or strata in the history of science (both as object and discipline) can also be related to the better-known fields of historical research in the English speaking world. Thus Imre Lakatos's work on the history of mathematics would be an obvious counterpart of Cavaillès, and his consciously anachronistic "rational reconstructions" of "research programmes" parallels Bachelard's "histoire sanctionée"; like Bachelard Lakatos wants to write the history of science "in the light of his rationality theory", i.e. Lakatos's objective is an internal, "*radically improved version*" of the history of science, a history that mentions "'false beliefs'" at most "in a footnote" (Lakatos (1978: 119). Laudan's evaluative scheme for research traditions, a scheme based on problem-solving power, is also naturally situated on this level. Finally, Crombie's work moving towards an anthropology of science seems to be concerned mostly with the level of epistemologized discourse; after all, he describes and explains how and when (new) disciplinary principles, styles of reasoning, images of knowledge and research mentalities emerge.

Be it noted also that Foucault's distinction between these various ways in which the history of science can – and should – be written sets him favorably apart from the somewhat imperialist tone of voice with which other methodologists present their own model to the exclusion of others.[27]

However, it would be a mistake to think that because for Foucault the archaeology of knowledge is concerned with just one possible stratum of historiography, its criticisms and methodological suggestions are of no avail to other approaches. Certainly the criticisms and suggestions of *The Archaeology* do carry over into other domains of the historiography of science: its criticism of the history of ideas stands and must be answered whether one adopts the archaeological edifice or not; its identification of anachronistic tendencies is irrelevant only for research programs that consciously adopt the viewpoint of the *histoire sanctionée*; its *Annalist* call for greater attention on the identification and delimitation of series holds for every historical project; its proposals for studying statements and textual strategies has not become obsolete due to recent work in discourse analysis; its suggestions on the interrelations between disciplines should be of interest even when studying discourses on the level of the epistemological or the scientific; and its refinement of the simple scheme 'prehistory vs. history' (of a science) might be relevant to historians and philosophers of science alike.

It is for this reason, i.e. the importance of the archaeological suggestions for *any* historiography of science, that there also is no need to follow those interpreters who construct something of a "break" between Foucault's archaeological and genealogical works. As I see it, Foucault could have left behind the general historiographical proposals of *The Archaeology* only by paying the high price of falling back to traditional, naive history. What genealogy amounts to then cannot be a rejection of the basic premisses of historical constructivism marked by the identification of series and the investigation into their correlations. Instead, genealogy implies the introduction of further strata. To put it in a nutshell, while *The Archaeology* is mainly concerned with the series of objects, modalities, concepts and strategies, while *The Archaeology* treats social factors more as conditions of these series rather than as series in their own right, genealogy analyzes more explicitly the series of social conditions themselves. To do this calls for definitions of new kinds of events, new kinds of relations, and new kinds of interactions between series and series of series. In short, it calls for a political theory of science, for a theory on the relations between power and knowledge.

NOTES TO PART I

[1] Sheridan (1980: 90): "The austerities - the aridities, some said - of *L'archéologie du savoir* could not fail to disappoint."

[2] For Merleau-Ponty, see Waldenfels (1983: 191); cf. Gutting (1989: 5). For Ricoeur, see Waldenfels (1983: 308).

[3] Cf. Gutting (1989: 5). Gutting does not give the reference to Cavaillès' text. The passage is in Cavaillès (1987: 76).

[4] For general background of the school, I rely on chapters nine and ten of Bourdé and Martin (1983). Cf. Stoianovich (1976).

[5] Furet (1985). A parallel reading of this text and Foucault's *Archaeology* is suggested in Bourdé and Martin (1983: 325).

[6] On Braudel, Ariès and Revel, see Revel (1986a); Althusser (1970a: 45); Canguilhem (1988: ix–x). For a more critical voice of an *Annalist*, see Vilar (1976: 13). See also Vilar (1985: 62–63).

[7] Even though the biological metaphor does not loom large in Lovejoy's writings.

[8] Above we already encountered the critique of members of the *Annales* school, of Bachelard, Canguilhem and Althusser. To this group of critics we can further add, among others; Michel Fichant (1969), Michel Pécheux (1969), Bruno Latour (1987, 1988), François Jacob (1974), Joseph Agassi (1963), Larry Laudan (1977), Conal Condren (1985), John Dunn (1968), W. F. Greenleaf (1966), J.G.A. Pocock (1972), and Quentin Skinner (1969).

[9] Interestingly enough, in writing about the *Foundations of Modern Political Theory*, Skinner himself deals more with ideologies and styles of legitimation than with the intentions in writing. (See Boucher 1985: 223.)

[10] Of course, this parallel should not come as a surprise; after all, the attempts of the first generation of *Annales* historians (Febvre, Bloch) to model the various mentalities of medieval and early modern times on the illogical and hypersensitive *mentalité primitive* of Lévy-Bruhl has been criticized even by members of the *Annales* School itself (Chartier 1982).

[11] This reformulation is influenced by Körner (1984: 118–19).

[12] I owe this example to Riitta Korhonen-Kusch.

[13] However, over and above an illuminating explication of the notion of "antidiscipline" little theoretical work has so far been done in this field (Lepenies 1978; Wilson 1977).

[14] In fairness to Foucault, I should mention that this observation is at least hinted at in the *Archaeology* in his distinction between different

thresholds.

[15] It is also inviting to ask to what system of modal logic these accessibility relations correspond. After all, different systems of modal logic, like *S4*, *S5*, *T*, and *B*, are semantically distinguished precisely by whether the accessibility relation between possible worlds is reflexive, and/or symmetric, and/or transitive (see, e.g. Hughes and Cresswell 1968: 62–81).

[16] Thus it turns out to correspond to the accessibility relation that characterizes system *T* of modal logic.

[17] Even a work like Thomas Kuhn's *The Structure of Scientific Revolutions* (Kuhn 1970) has had little impact on the historian's practice, even though some occasional critical remarks on Kuhn's scheme can be found here and there (see e.g. Reingold 1980).

[18] This is one of the central themes in Agassi (1963). See also Fichant and Pêcheux (1969).

[19] For more criticism of Cohen, see Agassi (1988a).

[20] Ludwik Fleck's *Entstehung und Entwicklung einer wissenschaftlichen Tatsache* (1935/1980) is perhaps the most important exception.

[21] For Popper's criticism, see Popper (1974). For Canguilhem's criticism, see Canguilhem (1988: 23). For Laudan, Toulmin and Hacking see below.

[22] Thus Paul Feyerabend has recently couched his radical antipositivism in terms of a comparison between the history of style periods and the history of science. (Feyerabend 1984). Already in the thirties, Ludwik Fleck, a later rediscovered early proponent of a similar model, spoke of "Denkstil" as a conglomerate of taken-for-evident judgments, methods and technical as well as literary devices. For Fleck, the style of thought prevailing within a period or a science puts constraints upon the ideas of its members, it "determines 'what cannot be thought otherwise'" (Fleck 1980: 130). Fleck's favorite example of a different style of thought that has become inaccessible to us were Paracelsus's writings (*ibid.*, 45).

[23] W. L. Wisan's work also deserves to be mentioned. He tries to preserve continuity in the history of science by shifting attention away from explicitly formulated theories towards "the emergence of new scientific styles" (Wisan 1981). Here a style is characterized by a "structure", which can be, e.g., classificatory or algebraic, by "content" as the subject matter, by "techniques" like philosophical analysis or geometrical methods, and by "expressive quality", e.g. "a feeling for the world of substances and essences" (*ibid.*, 325–28). Wisan suggests that whereas theories might emerge suddenly, the scientific styles upon which they are based evolve in

a slow and piecemeal fashion (*ibid.*, 312–14).

[24] The term "new experimentalism" was coined by Robert Ackermann. See Ackermann (1989).

[25] The table is constructed on the basis of Laudan *et al.* (1986).

[26] Yehuda Elkana's "images of knowledge" might also be situated on this level: "beliefs held about the task of science (understanding, prediction etc.), about the nature of truth (certain, probable, attainable, etc.), about sources of knowledge (by revelation, by ratiocination, by experiments through the sense)" (1981: 14).

[27] For instance, despite the parallels that exist between Foucault's and Quentin Skinner's criticisms of certain anachronistic tendencies in the history of ideas, Foucault's allowing for several registers in writing the history of thought enables him to avoid Skinner's absurd Baconist consequence according to which, as one critic puts it, "we can either become historians, and renounce the attempt to make connections with the present; or else, in Mr. Skinner's (strikingly Baconian) words, 'We must learn to do our own thinking for ourselves'" (Leslie 1970: 435).

PART II:

FOUCAULDIAN GENEALOGY

8. INTRODUCTION

As I have already pointed out in the introduction of this study, my reconstruction of Foucauldian genealogy is meant to be, by and large, independent of my reconstruction of the archaeology. This stated intention notwithstanding, and in order to provide the reader with some hints of what is to come, a few additional comments on the relations between the two projects are called for, nevertheless.

In his genealogical writings, Foucault's central interest does no longer lie with developing an apparatus for conceptualizing relations between statements and their indices. This type of series is not rejected as insufficient or unsatisfactory, but the focus is now upon defining a new set of series of entities and processes. These entities and processes are power mechanisms, power networks, interests, systems of exclusions and prohibitions, and coercive institutions. To be sure, the archaeological edifice does not rule out attention for such entities. Indeed, for a precise delimitation of series of objects, already *The Archaeology* calls for an investigation into institutions and social relations. Yet *The Archaeology* tends to treat such institutions and relations more as social conditions of discourses, rather than as series of entities and events in their own right. While in *The Archaeology* social conditions are invoked mainly as elements of explanations for the existence of certain forms of scientific discourses, genealogy raises the domain of the social to a level that itself stands in need of explanation and analysis. Put differently, genealogy seeks to provide a new conceptualization for social relations in general. Scientific discourse as a subject matter does not go by the board, but the set of principles of selection, the set of principles for constructing series of texts and discourses, is being enlarged. In Foucauldian genealogy, the existence of a discourse, at a given point in time, is only one of two possible phenomena to be explained. The new phenomenon attended to are relations and organizations in the social domain, and here the existence of discourses can itself be part of the explanation.

It is well in line with the archaeological-methodological demands that this new or additional principle of selection calls for some kind of theory as to what kind of entities or processes power mechanisms, power networks, etc. are, and how they are to be identified and studied.

Unfortunately though, Foucault's genealogy is a highly complex bundle of ideas, difficult to present coherently, and hard to assess. This is due partly to the fact that Foucault has not given us for genealogy, what he has provided us with for archaeology, i.e. a booksize elaboration of its main concepts, theses and methodological assumptions. Rather, his views have to be gathered from several historical monographs, his editions of historical source materials, numerous articles and – perhaps not surprisingly for a "French Mandarin" – also from dozens of interviews. Thus it is understandable that many writers in the field of science studies ignore or quickly bypass his contribution, and that studies of Foucault's genealogy either reflect the partial impenetrability of his texts or confine themselves to rather limited aspects dealt within them.

My discussion too will confine itself to only some facets of Foucault's later work. Ignoring Foucault's dislike for systematicity and stringency, I shall try to show that his various pronouncements on power and science can be reconstructed in a way that makes the different assumptions and parts of genealogy not only distinguishable but also separately open to assessment. In other words, Foucault's genealogy need not be the diffuse tangle of ideas which it easily seems to be upon first reading. I will also seek to show that the different theses of genealogy are often defensible either as they stand, or when supported by further arguments. To be sure, engaging in this reconstructive work frequently implies going beyond Foucault's own wording: occasionally I impose a different vocabulary upon his ideas, at several points I establish links between ideas of his and others that he himself did not attend to, and sometimes I search for new arguments that he might not have accepted himself. Obviously, this procedure can only lead to the construction of a 'Foucauldian genealogy', rather than to the faithful unfolding of 'Foucault's' genealogy.

I shall start from a reconstruction and defense of Foucault's *theory* of power. For want of a better term, I propose grouping under this heading of 'theory' those of Foucault's pronouncements that suggest what constitutes a power relation, how power interacts with knowledge, what types of power can be distinguished, and how power works. For analytical purposes, this kind of *theory* of power might be distinguished from three other ventures *vis-à-vis* power, to wit, a *methodology* of how power and its effects are to be investigated, a *political criticism* of vicious forms of power, and a *historical analysis* of specific forms of power in different historical places and periods.

In my discussion of the first two ventures, I shall occasionally draw attention to parallels and differences between Foucauldian genealogy and recent work in the sociology of science (Barnes, Bloor, Collins, Latour, Schaffer, Shapin). On the one hand, the parallels between Foucault and the sociologists of science are noteworthy because even where Foucault does reach conclusions strongly reminiscient of the sociology of science, he arrives at them via a different route. Thus his independent arrival at similar views might be looked upon as providing independent support for certain theses of the sociologists. On the other hand, the differences between Foucault and the sociologists are interesting since they might lead to advancing our understanding of what a social reconstruction of the history of science is or should be. (E.g. what role a theory of power should play in social studies of science.)

Finally, in a brief, concluding, chapter, I shall defend a Foucauldian strategy for political criticism of science, a strategy that stresses the importance of questioning rationalities and that abstains from basing itself upon an elaborate normative, or utopian theory of science or society.

9. THE CONCEPT OF POWER

My treatment of what I propose to call Foucault's theory of power is structured around a distinction first advanced with respect to justice (Rawls 1972: 5–6) and later applied to power (Lukes 1974: 26–27). This is the distinction between *the concept* and *conceptions* of power, where *the concept* of power refers to the basic core or primitive notion of power lying behind all (or most) talk of power in the social sciences and philosophy, a basic core that is developed or fleshed out by different authors into distinct theoretical *conceptions*.

Even though I do not believe that there is one single concept of power underlying all theories of power, I shall try to show that at least the best-known sociological work on the notion of power (Lukes 1974), on the one hand, and Foucault, on the other hand, roughly share the same concept of power. In establishing this point, I hope to convince the reader that, when speaking of power, Foucault and Lukes, are not referring to totally different phenomena.

It is only subsequently that we shall go into the details of Foucault's *conception* of power. As we shall see, this conception is marked not only by its employment of – partly incompatible – *models* of power, but also, for instance, by its stress on mechanisms of power, its emphasis upon the relations between social power and scientific knowledge, and its idea of power as an omnipresent element of social life.

Some Standard Distinctions

Let us start by recalling, somewhat telegraphically, some of the main notions in the standard literature on power, namely, the counterfactual analysis of power, the different constituents and aspects of power relations, and the various forms or 'species' of power. (The reader familiar with these notions can proceed directly to the next section.)

We attribute power not only to people – individuals or collectives – but also to animals, machines, God(s), and abstract things. We say, for instance, that boxers have more power in their fists than most other men, and we speak of the power of teachers over their students, of the power of a crane, or the power of love. Even though it is hardly informative, we might say that all these attributions are attributions of capacity to bring about or hinder the coming about of states of affairs or processes (Barnes 1988: 2).

Contrasting the power of a crane and the power of a boxer's fists with the power of a teacher suggests a first distinction, i.e. the distinction between social and physical power. Of course, a boxer's physical power is connected in various ways to his social power, and even the physical power of a crane can be involved in all kinds of social relations. Still, the ability, power, or capacity to do physical work can be set apart from the ability to get someone to do, or abstain from doing, something.

However, 'getting someone to do, or abstain from doing, something' is still an insufficient characterization of what we mean by social power. This observation has led writers on power to conceptualize power relations in terms of counterfactuals and interests. The underlying intuition seems clear enough. Were it not for the actions of the one who "has" or "exercises" power, i.e. the actions of the "power holder", the victim, or "power subject", would have acted otherwise, that is, in a way conceived by her to be more advantageous, more in accord with her own interests.[1]

However, linking the attribution of power to the possibility of the power subject to have acted otherwise and linking it to her interests is not without problems. In practice it might often be rather difficult to determine just how the power subject *would have* acted, or what her *interests* are. While admitting these problems, Stephen Lukes, the writer whose views have perhaps been most widely discussed in recent years, has suggested that these problems can nevertheless be overcome. Historical knowledge on how people act under various circumstances and a notion of real interests, i.e. interests that agents do not have to be aware of themselves, allow us, or so Lukes alleges, to formulate the relevant counterfactuals even in cases where *prima facie* we have no direct knowledge of the deliberations of the power subjects' themselves (Lukes 1974: 46–50).

However, to analyze relations of power does not mean to study only the interests of the power subjects. The study of power is also concerned with distinguishing between different constituents of power relations, and with driving conceptual wedges between different forms of power.

Different authors have suggested and developed various distinctions between different constituents of power relations, but it seems fair to say that the following list has emerged as something like the standard view on this topic[2]:

(a) the *base* of a power relation refers to the resources (e.g. economic wealth or prestige) that the power holder can draw on in order to influence the actions of power subjects';
(b) the *means* of the power relation are the specific actions (e.g. commands or threats) the power holder employs in the attempt to subdue the power subject;

(c) the *scope* of power is the set of actions that the power holder, by employing her means and resources, can get the power subject to perform or abstain from performing;
(d) the *amount* of power denotes the net increase in probability of the power subject's performing (or abstaining from performing) a specific action due to the actions of the power holder;
(e) the *extension* of power is the set of individuals over which a given power holder has power;
(f) the *costs* of a power relation are the various costs that arise for the power holder from establishing and maintaining the power relation;
(g) the *strength* of a power relation refers to the various costs that arise for the power subject in case she refuses to enter and remain within the relation of power; and
(h) the *intensity* or *zone of acceptance* of a power relation is the set of actions which the power holder can get the power subject to perform without loss of the latter's compliance.

Turning from constituents to forms of power relations, philosophers and sociologists have much debated the relations between the notions of authority, coercion, deterrence, force and manipulation. While different writers disagree as to how precisely we are to draw the lines between these concepts, they generally agree roughly as to the completeness of this list. There also is no general agreement as to whether force or violence should fall under the concept of power. Authors who answer negatively draw on the intuition that to be a victim of a power exercise seems to leave us with some - however small - possibility of choice between compliance and non-compliance, a possibility absent in the case of violence.[3] This is not to deny, of course, that violence and power can be, and often are, tightly coupled; for instance there are cases where violence figures as the ultimate threat by means of which the power holder upholds his power. It is in part because of this link, in part because of a reliance on a broader definition of power as "the capacity of some persons to produce intended and foreseen effects on others" that some authors see force as one species of (social) power rather than as a distinct category.[4]

Disregarding the fact that different writers disagree over details, let us briefly summarize, for later reference, one way of distinguishing between forms of power. According to Lukes (1974: 32), coercion and force involve open and observable conflicts of interest between power holder and power subject. Manipulation also presupposes a conflict of interests, but in its case the conflict of interests is not perceived by the power subject. Authority is many-facetted in that we tend to attribute it not only in cases where we observe or assume a conflict of interests but also where we do not even assume any conflict of interests at all. In general, we speak of authority where the power subject conceives – at least in the long-run – of the power holder's demands as rationally grounded and/or as beneficial to herself as the power subject. The following picture summarizes these distinctions (Figure 12).

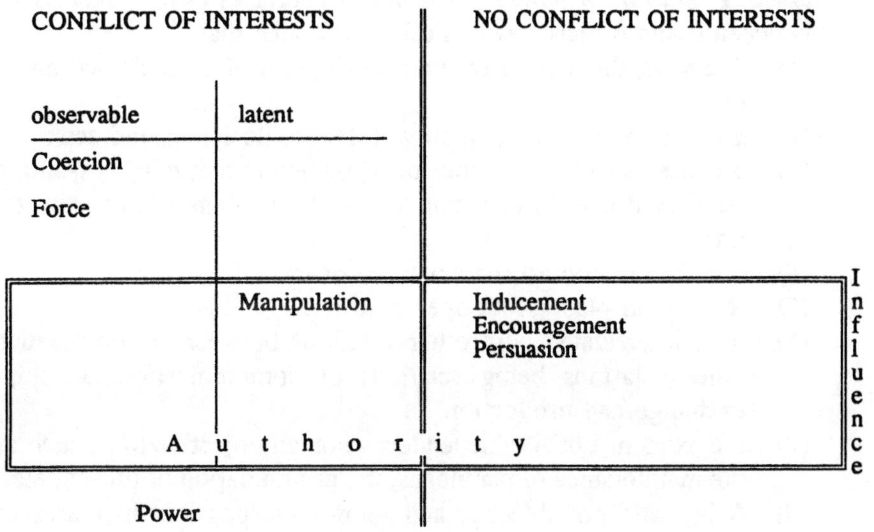

Figure 12

A FOUCAULDIAN DEFINITION OF POWER

With some of the more central notions and distinctions of more traditional study of power recalled, we can return to Foucault. Relying mostly on his 1982 article "The Subject and Power", the following 'Foucauldian definition of power' can be formulated. This thirteen-part definition can serve us as a first brief overview of our author's conception of power. More specifically, this definition allows me to substantiate my claim concerning a concept of power found both in Foucault and Lukes, and to highlight, in an anticipatory fashion, some central features of Foucault's theory.

> Between two individuals (or group of individuals), *a* and *b*, there exists a *relation of power* (such that *a* exercises power over *b*) iff between *a* and *b* there exists a relation *R* such that
>
> (*1*) due to *R*, the actions of *a* reorder the set of possible actions of *b*;
> (*2*) *a* recognizes *b* as a person who makes decisions and acts;
> (*3*) *b* consents to the existence of *R*, believing that *a* has a winning strategy if she (*b*) does not comply to *a*'s demands or expectations;
> (*4*) *b* is free to contest the existence of *R*;
> (*5*) *R* holds in place of an open conflict;
> (*6*) *R* is interrelated with other relations between *a* and *b*, such other relations being relations of communication, kinship, exchange and production;
> (*7*) the existence of *R* is dependent upon an objective of *a*, such as the maintenance of privileges, the accumulation of profits, etc.;
> (*8*) *R* is made possible by, and reinforces (possibly increases or modifies), a system of differentiation (determined by law, tradition, status, privilege, economic differences, linguistic or cultural differences, competence) operating between *a* and *b*;
> (*9*) the existence of *R* is dependent upon specific means at *a*'s disposal, means that bring the power relation into being, and uphold it (threat of arms, effects of words, economic disparity, means of control, surveillance, rules);
> (*10*) *R* has a specific form or degree of institutionalization;

(*11*) *R* is rationalized to some degree (effective, certain in its results, related to, and productive of, knowledge);
(*12*) the existence of *R* allows for changes with respect to conditions *7–11*, and it is related to further relations of power (*R'*, *R"*, ...) that hold between *a* and *b* (or between *a* and other individuals, or between *b* and other individuals);
(*13*) The identity of *a* and *b* is (at least in part) determined by *R*.

Especially theses (*1*) to (*4*), (*8*), (*9*), and (*11*) of the Foucauldian definition parallel some of those ideas referred to in the beginning of the present chapter.

As concerns (*1*), Foucault says that power "is a way in which certain actions modify others ..." (S&P 219). To exercise power is "to structure the possible field of actions of others" (S&P 221). Two points are worth noting here. On the one hand, the stress on actions is meant to distinguish power from violence, while, on the other hand, the notion of structuring – or rather re-structuring – of actions points to a counterfactual analysis of power.

Foucault draws the distinction between power and violence in a pretty standard fashion. Violence or force leave the victim no choice between compliance or non-compliance; as Foucault puts it, in cases of violence the victim is treated as a body and not as a decision-maker, i.e. as a physical object rather than as a person whose actions are to be influenced:

> [power] ... is a mode of action which does not act directly and immediately on others. Instead it acts upon their actions: an action upon an action ... A relationship of violence acts upon a body or upon things (S&P 220).

That Foucault accepts a counterfactual analysis of power is not only hinted at in his notion of a "structuring of actions"; he also expresses this idea more straightforwardly in the following passage: "[if e.g. a man] can be induced to speak, when his ultimate recourse could have been to hold his tongue ... [then his] freedom has been subjected to power" (PPHC 83). In other words, speaking of power presupposes that, were it not for the action of the power holder, the power subject would or could have acted otherwise.

Condition (*2*), i.e. the recognition of *b* by *a* as someone who acts and makes decisions, is naturally seen as a corollary of this distinction between violence and power. In the case of a power relation one of the actors tries to get another *to do* something, tries to get another *to act* (or, to abstain from acting). In Foucault's words, it is an "indispensable" element of the power relation that "'the other' (...) be thoroughly recognized and maintained to the very end as a person who acts" (S&P 220).

Ingredients (*3*) to (*5*) can be taken up jointly. Again, Foucault's suggestions are not only well in line with mainstream sociology, but they are also further elaborations of (*1*) and (*2*). Since in the case of power – in counterdistinction to the case of violence – the victim is involved as an actor and not just as a body, he or she remains able to choose between either compliance or non-compliance: "Power is exercised only over free subjects, and only insofar as they are free ... slavery is not a power relationship when a man is in chains" (S&P 221). The last point is interesting, because in claiming that the slave is no longer free, Foucault distances himself from Sartre who once suggested that the radical freedom of the human being extends even to the slave in chains, in so far as the latter still has to decide what his chains mean to him (Sartre 1969: 483–84). Instead, Foucault proposes that a power relationship demands the possibility of "refusal or revolt", even when the latter might consist only in preferring death over submitting (PPHC 83). In thus stressing that "resistance is never in a position of exteriority in relation to power" (*ibid.*), Foucault comes of course close to restating Max Weber's well-known suggestion according to which power is the probability of individuals realizing their wills *despite the resistance of others.*[5]

Furthermore, Foucault puts much emphasis on the observation that a power relation exists always *in place of* an open conflict (*5*) and that the submission of the power victim is due to the latter's belief in her lack of a winning strategy for such a conflict (*3*).

More precisely, Foucault sees the existence of a power relation as an unstable equilibrium of conflicting interests in both the power holder and the power subject. It is the inevitable instability of power relations which justifies the idea that "at every moment the relationship of power may become a confrontation between two adversaries" (S&P 226). Looked at from the side of the power holder, the instability can be expressed by saying that the very existence of the power relation is an indication of limitations on her part. While, ideally, the power holder would like to do away with the power subject's freedom altogether – thus replacing the relation of power by one of violence – the very existence of the power relation is *the* indication of her inability to obtain this most favorable result. As long as the power relations holds, it is always possible for the power subject to revolt and resist, i.e. to engage in a struggle costly for both sides (even where the power holder is likely to win). In other words, as long as the power relation obtains, the power holder is forced to accept her victim as a decision-maker and actor. The power subject, on the other hand, is faced with the alternatives either of entering into a confrontation she feels will be costly – at least given the price of the demanded action or non-action – or of yielding to an uncomfortable or unpleasant, perhaps humiliating, demand.

In "The Subject and Power", Foucault stresses with respect to condition (*6*) that relations of power are interwoven with, but distinct from, relations of communication between the power holder and the power subject. He also suggests that relations of power between human beings are to be distinguished from the 'power' of natural objects or machines, or 'power to', in general. Like other authors before him, Foucault refers to the latter relations as relations of "capacity":

> Power relations, relationships of communication, objective capacities should not ... be confused. ... This is a question of three types of relationships which in fact always overlap one another, support one another reciprocally, and use each other mutually as means to an end (S&P 217).

References to the fact that relations of power are interwoven with economic, kinship and sexual relations can be found in other places of Foucault's oeuvre (e.g. P/K 142). As we shall see below in greater detail, it is Foucault's aim to defend the genuineness of power relations, that is, he seeks to argue foremost that relations of power cannot be reduced to economic relations.

As concerns (*7*), i.e. "the types of objectives pursued by those who act upon the actions of others" (S&P 223), suffice it to say here, for the purpose of a first overview, that even though Foucault thinks of power relations as being due to the objectives of power holders, he does not wish to maintain that all relations of power and *all effects* of power are intended or foreseen. Indeed, the allowing for some kinds of invisible hand or counterfinality phenomena is central to a Foucauldian theory of power. Later we shall return to this point in greater detail.

Foucault's "system of differentiation" (*8*) and his "means of bringing power relations into being" (*9*) (S&P 223) need little explanation at this point. Obviously, these systems and means correspond fairly straightforwardly to what the received view refers to as "bases" or "means" of power.

Conditions (*10*) to (*13*) are of the utmost importance to Foucault's conception of power. (*10*) captures Foucault's idea that relations of power never exist in a social vacuum. Power relations not only presuppose institutions like e.g. the family, the school, the university, the police, but they also uphold these institutions and make them possible in the first place:

> ... the great strategies of power [i.e. the systems of institutions] encrust themselves and depend for their conditions of exercise on the level of the micro-relations of power. But there are always also movements in the opposite direction ... (P/K 199–200).

(*11*) is concerned with the essential relation between knowledge, power and costs. The means by which a relation of power is upheld can be more or less effective, i.e. more or less costly to the power holder. The latter will seek to keep the costs of control as low as possible by constantly increasing the effectiveness of these means (S&P 223–24). This is one of the several ways in which power and knowledge are interconnected, ways that will later occupy our attention at some length.

(*12*) captures Foucault's constant reminder that power relations form "networks", that they "are superimposed, ... cross, ... sometimes cancel one another out, sometimes reinforce one another" (P/K 99). Put differently, "relations of power-knowledge ... are matrices of transformations" (HS I 99).

Finally, (*13*) refers to the idea that our identity as individuals cannot be separated from the power relations in which we live and act; it is in and through power that "human beings are made subjects" (S&P 208).

10. THE GENEALOGICAL CONCEPTION OF POWER I: FIELDS AND NETWORKS

Models of Power

In the last chapter I attended to parallels between Foucault and the received analysis of power only insofar as the basic skeleton or core assumptions were concerned. Further parallels can be drawn and will be taken up below at least at those points where such allusions can serve to illuminate or strengthen the Foucauldian position. In any case, even the parallels referred to up to this point suffice to show that Stephen Lukes's assessment of Foucault's theory of power as radically different from Lukes's own tradition, more precisely, as "diffuse" and as unable to allow for any links between the attribution of power and the attribution of "causal or moral ... responsibility", is exaggerated at best (Lukes 1986a: 15).[6]

Turning from parallels to differences, be it noted first that Foucault shows little interest in distinguishing between different "species" of power. This is due to the fact that he is mainly concerned with *mechanisms* of power exercises, a concern that, as we shall see, cuts across various forms of power. This is not to say, however, that the reconstructed Foucauldian definition of a power relation cannot be related to the distinction between various "species" of power, even though this attempt calls for some refinements. While certainly most of the ingredients of the Foucauldian definition hold for coercion and authority alike, conditions (*3*) to (*5*) obviously cannot apply in the case of manipulation. This is because manipulation, like force, does not allow for a decision over compliance or non-compliance by the power subject. Nevertheless, there can be no doubt that Foucault takes manipulation to be a form of power. Perhaps the most natural way to resolve this problem is to think of the above definition as defining the paradigmatic case of power, i.e. coercion, and as being extendable to cases like manipulation and force by relaxing or dropping some of the thirteen conditions.

A second feature that sets Foucault apart from the better known analyses of power is his 'models of power', i.e. his use of various, divergent, and in some respects even contradictory conceptualizations of power. It is hard to decide to what degree the employment of such models of power is the result of a systematic and deliberate choice or the result of conceptual confusion and vagueness. Thus one cannot refute Nancy Fraser, perhaps Foucault's most perceptive critic, when she charges him with ending up "with a curious amalgam of amoral militaristic description, Marxian jargon, and Kantian morality" (Fraser 1981: 284; cf. Fraser 1985). Actually, one can even add to Fraser's list insofar as Foucault occasionally also uses the terminology of physics to characterize aspects of power: he does not only call his analyses a "microphysics" of power, but his references to "forces" also often have overtones more of physical field theories of force rather than of military action.

Yet even if one is dissatisfied with the specific manner and degree of sophistication with which Foucault employs his models, and even if one disagrees with his choices for models from the start, it still seems possible and reasonable to defend the use of a plurality of metaphors, analogies and vocabularies (and even conflicting ones) in the study of power.

A parallel with developments in the philosophy of physics provides support for such a position - and the fact that it does so already in part justifies calling Foucault's conception of power a "microphysics". Nancy Cartwright (1983) and Ian Hacking (1983) have stressed the importance of the use of a variety of - often mutually inconsistent - models in modern physical theories, e.g. in quantum mechanics. Physicists usually work with several models for different purposes, models that often are not deducible from their common general theory. Taking these and related observations on the character of physical theory as their starting point, Cartwright and Hacking deny that the most general laws of physics can intelligibly be regarded as true. What can meaningfully be regarded as true are rather the very local, simple laws that are bound to a given model. Cartwright, with her tongue in her cheek, concludes that God cannot possibly have the elegant, unifying mind of a French mathematician, but that He necessarily must have the untidy mind of an Englishman (1983: 19). Hacking suggests that this God is not very far from the Leibnizian God who creates a maximum of phenomena governed by the greatest possible simplicity of laws: "the best way to maximize phenomena and have the simplest laws is to have the laws inconsistent with each other, each applying to this or that but none applying to all" (1983: 219).

The way one can transfer these ideas to social science and thus to a defense of Foucault is obvious enough. If social reality was not designed by a French mathematician, then it just will not do to claim that either game theory or Marxist notions or theories of communication provide the one and only key to the study of power. To assume that social reality – and the realm of power relations that is coextensive with it – is at least as messy as the subject matter of physics, leaves us no choice but to come to grips with its various, often conflicting features in terms of various, often incompatible, models. This is admitted even by Stephen Lukes, a writer on power who strives for simplicity and stringency in his own theory. Lukes writes that "what unites the various views of power is too thin and formal to provide a generally satisfying definition applicable to all cases" (1986: 4–5).

To repeat, this defense of a plurality of models in a theory of power does not imply a defense of the Foucauldian models. Let it be said, though, that there certainly are no apriori arguments against his models either. There might well be insights to be gained from looking upon the social body as a Faradayan field of forces, thereby assuming forces – alias powers – to be primary with respect to physical bodies – alias individuals. And the model of war – e.g. in a Clausewitzian conceptualization – might be fruitful in coming to grips with certain features of the political or scientific struggle.

It is true that these models are at least partially incompatible with the standard conceptualization of power in terms of agency and responsibility, a conceptualization that as we saw above also appears in Foucault's own definition of power. Yet Foucauldians have at least two possible ways of answering this observation. On the one hand, they might hold that even though all power is analytically linked to agency, we might still gain new insights into power by bracketing this link at least occasionally. On the other hand, they might argue that although there are forms of power where the link between power and agency holds, there are still other forms, effects and structures of power that have little or nothing to do with either agency or responsibility. On this latter view one thus has a double reason for employing agency-excluding models: for illuminating agency-involving power structures from a new angle, and for conceptualizing phenomena of power where agency has no role to play.

Foucault's use of these various tools becomes more transparent as we turn to *key theses* of his theory of power, theses that constitute the basis of his conception of power. These theses concern the nature of a power relation, the omnipresence of power, and the distinction between micro- and macropower. Discussing these theses will pave the way towards a later defense of Foucault's view on the relation between social power and scientific knowledge.

Internal Relations

To start with the nature of the relation of power, note that Foucault stresses time and again that we should not look upon power as being *possessed* by some individuals or groups but as being *exercised* by some individuals over other individuals:

> Power is not something that is acquired, seized, or shared, something that one holds on to or allows to slip away; power is exercised (HS I 94).
>
> [...] power is neither given, nor exchanged, nor recovered, but rather exercised, and [...] it only exists in action (P/K 89).

We only need to add to these quotations Foucault's further remark to the effect that a relation of power must be taken as "strictly relational" (HS I 95) to see how naturally Foucault's claim is rephrased in terms of the classical Leibnizian distinction between "external relations" (of comparison) and "internal relations" (of interaction).[7] *A power relation*, for Foucault, *is an internal relation of interaction, not an external relation of comparison*.[8]

Recall that for Leibniz the distinction in question is one that is based on the criterion of the reducibility of relations to one-place predicates of the *relata*. For example, "Titius is wiser than Caius" can be reduced to the two sentences "Caius is somewhat wise" and "Titius is very wise" by relying on the general truths according to which "wiser" equals "superior in point of wisdom", and "'very' represents a degree superior to 'somewhat'". Because of the possibility of this reduction, "Titius is wiser than Caius" qualifies as an external relation of comparison (*relatio comparationis*). However, "Paris loves Helen" cannot be reduced in this way. This is because in this sentence Paris does not only have the feature of "Being a lover", but also the feature of "Being a lover in virtue of – or insofar as, or since (*propter* or *eo ipso*) – Helen being the beloved". In other words, a reduction to one-place predicates is not possible in this case, since "Being the lover of someone" is still a *relational* property. Thus "Paris loves Helen" is a relation of connection (*relatio connexionis*), i.e. an internal relation of interaction.

Attributing to Foucault the conception of power relations as internal enables us to explain his emphasis on "mechanisms" of power, his pronounced "anti-economism" and his frequent reference to metaphors of war.[9]

Because Foucault is above all interested in power relations as forms of interaction, i.e. because he centers attention on the exercise of power, he focusses less on the *relata* of that relation than on the *relatio* itself. That is to say, he concerns himself with the mechanisms (of control, coercion, manipulation, etc.) that figure in power relations:

> History has studied those who held power [...]; contrasted with this there has been the history of economic processes and infrastructures. [...] But power in [...] its mechanisms has never been studied (P/K 51).

Thus Foucault does not confine himself to conceptualizing the power relation (*R*) as an act issuing from the power exerciser (*a*) to the power subject (*b*):

$$a \xrightarrow{R} b$$

... but he rather proposes that the power relation be conceptualized 'inside-out'. In other words, the mechanisms necessarily involved in relations of power are taken as determining *a* and *b*, i.e. as shaping *a* and *b* in different ways. This implies that we must attend not only to *a* and *b*, but also to their respective relations (*R'*, *R"*) to the mechanism (*m*) involved in *R*:

$$\underset{R(a,b)}{a \xleftarrow{\;R'\;} m \xrightarrow{\;R''\;} b}$$

At this point I shall not go into the details as to what kinds of mechanisms Foucault has in mind here.[10] Be it mentioned here only that *Discipline and Punish* deals, for example, with what one might call mechanisms of "ecological control"[11], that is, mechanisms for limiting and pre-structuring the spatial environment, temporal limits, and social *entourage* of the actions of individuals. Thus in the prison or the workplace, individuals are confined – *inter alia* by architectural measures – to limited spaces, trained to carry out actions in specific and meticulously planned time intervals, and placed in ways so as to be under constant surveillance. Going beyond Foucault's own interests we might of course also think of even more down-to-earth mechanisms, e.g. the power mechanisms involved in conversation, such as interrupting or questioning, mechanisms much attended to in recent linguistic pragmatics.[12]

For Foucault, the choice between a conception of power as possession and a conception of power as interaction is closely connected to the alternative of studying power via economic or via military models. Indeed, Foucault attacks what he regards as the "economism of the theory of power" (P/K 88). Foucault suggests that both classical political theory and Marxism are representatives of an economism in the theory of power.

In the classical political theory, economism is closely tied to the traditional preoccupation of power theorists with questions of right, law and legitimization. This link, Foucault holds (P/K 88–90), can be seen especially clearly in the principal idea of contractualism according to which presocietal individuals 'trade in' their powers for protection provided by the state. Power is thus conceptualized as a commodity. On the premisses of this model, i.e. "the juridical" or "liberal" model of power, critical questions concerning, and analysis of, power, remain confined to a reflecting on the conditions of the initial or ideal contract, that is, confined to ever new justifications of either prevailing or imagined bodies of law in terms of the initial trading-in of pre-societal, individual powers. It is one of Foucault's most important claims in his historical work that this model of power with its inevitable conceptual link between power and law is impotent when applied to the study of modern mechanisms of power. This is because these mechanisms – that derive from medicine, the prison, and the army – operate not on the axis of just/unjust or legitimate/illegitimate, but rather on the axis of normal/abnormal or healthy/sick.

In Marxism the economism figures somewhat differently. Even though power is here taken as an internal relation and is not conceptualized as a commodity, its interactional aspect remains underdeveloped since Marxists are preoccupied with the allegedly "ultimate" causes of power in economic structures rather than with the actual effects of power mechanisms upon humans:

> On this view, then, the historical *raison d'etre* of political power is to be found in the economy. ... Is it [i.e. power] always in the service of ... the economy? Or, on the contrary, do we need to employ varying tools in its analysis ...? ... If that is the case, it is not the models of functional subordination or formal isomorphism that will characterize the interconnection between politics and the economy (P/K 89).

Foucault proposes "Nietzsche's hypothesis" as the most promising successor to an economically-minded study of power. According to "Nietzsche's hypothesis", "the relationship of power lies in the hostile engagement of forces" and is to be analyzed "primarily in terms of *struggle*, *conflict*, and *war*" (P/K 91).

"Nietzsche's hypothesis" clearly brings out the interactional aspect of power. Recall only that for Clausewitz in his classic *Vom Kriege*, war is "an extended duel", "always the clashing of two living forces", or "an act of violence in order to force the opponent to fulfill our wishes" (Clausewitz 1918: 3, 6). Foucault goes beyond Clausewitz, however, by inverting the latter's notorious *dictum* that war is but the continuation of politics by other means (P/K 90–92). To put the point more sharply than

Foucault does himself, whereas Clausewitz claimed that politics is the underlying deep structure, or "logic", of the surface structures or "grammars" of both diplomacy and warfare (1918: 641), Foucault holds that it is the *raison de la guerre* that underlies peacetime politics as much as military clashes: "Even when one writes the history of peace ... it is always the history of [...] war that one is writing" (P/K 91).

In fact Foucault claims that even Clausewitz's own *dictum* was already the inversion of an earlier principle of political thought. Interestingly enough, Foucault claims that the proponents of that earlier principle of politics as the continuation of war by other means[13] all mark a countermovement to the long and still dominant liberal conception of power. For them, war is never over, "law emerges from burnt cities and devastated countries", society is always divided in two army-like sections, and all theoretical discourse is but part of this ongoing battle (LK 11–13).[14]

Power and Personal Identity

The fact that Foucault focusses on the question of what power relations or mechanisms make of us, rather than on why individuals coerce or manipulate one another, implies the possibility of a second way in which the distinction between internal and external relations can be brought to bear on an attempt to reconstruct his theory of power. The crucial point of this second application is that the classical dichotomy of internal *versus* external relations was usually connected to the opposition between essential and accidental properties (Elster 1978: 22). Internal relations were generally regarded as essential, and external relations were conceived of as accidental; in other words, internal relations but not external relations were treated as constitutive of what their relata *are*.

Applying this dichotomy between internal-essential and external-accidental relations to relations of power, the second of Foucault's key theses can be formulated as stating that *relations of power are internal-essential rather than external-accidental*. That is to say, insofar as power relations make us what we *are*, and insofar as relations of power are constitutive of us as subjects or individuals, we cannot define ourselves, nor understand who we are, without understanding the mechanisms of power that shape us, without appreciating and analyzing our ability or tendency to take up specific roles within power relations.

In line with this reconstruction, Foucault calls for the study of how "peripheral *subjects*" are constituted "as a result of the effects of power" (NGH 98). And in another passage we read:

> [...] we should try to discover how it is that subjects are gradually, progressively, really and materially constituted through a multiplicity of organisms, forces [i.e. powers], energies, materials, desires, thoughts, etc. (P/K 97).

In still other terms, the central question is "how are we constituted as subjects who exercise or submit to power relations?" (ENL 48).

In his historical studies on the prison, the asylum and the production of the modern conceptions of health and sexuality, Foucault provides plenty of material for dealing with this question in terms of *historical analysis*. Here, however, I shall confine myself to some comments from a more topical perspective.

To start us off, note that Foucault's remark, according to which his study of power is to be a "microphysics of power" (DP 29), can - to some extent at least - be justified at this point. After all, modern physics (microphysics or quantum mechanics and the theory of relativity) differs from classical Newtonian physics most impressively in its abandoning of the substance vs. attribute scheme, in its giving predominance to relations and structures rather than objects, and in being a science of effects rather than a science of objects. For instance, whereas in the 19th century it was still thought that science starts from "real" objects and subsequently proceeds to conjectures as to their relations, microphysics gives priority to relations and treats "real" everyday objects as mere phenomena. The development of the theory of relativity has produced a way of thinking that conceives of relations as independent of their termini, and that conceptualizes objects "as strange functions of the function that relates them" (Bachelard 1971: 34). The particle in microphysics is no longer a classically conceived "extremely small body"; rather, the particle has been stripped of almost all of those determinations that classical physics attributed to physical bodies. The protons and electrons of modern physics are not material objects with the property of electricity because we know of electricity only as electrons and protons: substance and attribute can no longer be kept apart. Moreover, the particle in quantum mechanics has neither a definite absolute dimension, nor a precise location. One speaks no more about its size but about its "zone of influence" or "its power of opposition", i.e. the distance one electron can approach another when fired at it with a speed close to that of light (*ibid.*, 62).

Perhaps it was these characterizations of modern physics, characterizations that loom large in Bachelard's epistemological writings, that prompted Foucault to speak of his investigations as a "microphysics of power". After all, Foucault, too, argues for the priority of relations (of power) over individuals: he denies that power can ever be "localized here or there" (P/K 98), and he tries to show that the individual is the result of forces or powers, powers that – as we shall see later – cannot be modelled upon the power relations between larger social, "macrophysical", units.

Obviously notions of physical theory that resemble Foucault's conception of power predate the twentieth century. Foucault's own talk of "fields of forces" also invites reference to Faraday's conception of fields of force with its revolutionary idea of forces as being primary with respect to bodies.[15] Replace 'bodies' with 'subjects' and the analogy works indeed. Furthermore, Faraday also sounds Foucauldian, and Foucault Faradayan, because of Faraday's view that the amount and the direction of force associated with each point in the field of force is dependent upon, and determined by, the amount and the direction of force in contiguous points. Foucault conceives similarly of the social field of society. Not only are the power or force relations essential to the points of the field, i.e. the individuals, but he also holds that none of them can be described and explained without taking into account its power or force relations to others. Even Faraday's novel notion of "action in the vicinity" or "near-action"[16] has a parallel in Foucault's preoccupation with ecological mechanisms of power.

Turning from microphysics to objections, the major threat for the conception of power relations as internal-essential might be seen to arise from the observation that if *all* power relations in which we have been, and in which we live now, are essential to our identity, then the hope of us ever being free of those power relations must inevitably be vain: what is essential to our identity cannot be made to disappear without that identity disappearing, too.

It certainly makes no sense to claim that all power relation *tokens* – relations in which we figure as *relata* – are essential to what we are. The fact that we let ourselves be coerced into drinking a second beer by a friend who uses some trivial, minor threat, can hardly as such be part of what we are. Yet while it is unnatural to say that all of our power relations are essential to our identity, it is still plausible to assume that more lasting power relations, as well as our tendency and/or ability to enter certain types of power relations as, say, coercer or coerced, do have an intimate relation to what we are. In other words, some of our more lasting power relations and tendencies of reacting to power relations are "productive": they shape us into what we are.

For instance, the power relations and mechanisms that we have established with our students, spouses, long-term friends and colleagues, parents and teachers, seem to be likely candidates for having to be included into the rough definition of what we are, for having to be included into the set of determinants of how we understand ourselves.

The case of tendencies and abilities is no less clear. Our ability to act as, say, coercers, or our tendency to figure as likely victims of coercion, does not stand isolated among our personality traits. Obviously, since at least some of our more permanent traits are definitive of what we *are*, and since personality traits predisposing us to being either good or bad coercers can hardly be excluded from these central traits, our tendency and/or ability to act as power exerciser or power subject is an essential part of our identity.

It is true that, in some sense, the thesis of power relations as internal-essential, forces us to say that *we* do not remain what we are (or were) when essential power relations, or tendencies to take up specific positions in them, change. Yet rather than being a threat to Foucault's view of power relations as essential, this implication might also be seen as harmonizing with recent suggestions concerning the relativity of personal identity.[17] We all change over time, and over longer periods of time small-scale changes add up to bigger ones, so much so that it is reasonable to say that different time-slices of "our lives" are like different persons. If this idea is accepted nothing seems more natural than to extend it explicitly to power relations, too: when ridding ourselves of long-term and more fundamental power relations, or when becoming entrapped in them, and when either developing, or freeing ourselves from, tendencies to take up specific positions in specific types of power relations, we become different, new persons.

Networks of Power

As already indicated above, Faraday's idea that the force associated with each point influences, and is influenced by, the forces associated with its adjacent points, has a counterpart in Foucault. This counterpart is Foucault's claim that every individual both exercises power and is the target of the exercise of power by others. Just as for Faraday in the field of force there is no point without force, and thus no point without relations to other points, so for Foucault in the social field there is no individual without his or her relations of power to some other individuals. This claim can be regarded as the third key assumption of Foucault's theory of power: *Power is an omnipresent network in the social field*.

Foucault holds that "power is everywhere [...] because it comes from everywhere" (P/K 93) and he also writes that ...

> between every point of a social body, between a man and a woman, between the members of a family, between a master and his pupil, between every one who knows and every one who does not, there exist relations of power ... (P/K 187).

Reminiscent of Faraday's notion of "networks of lines of force" and his idea that forces can travel through a field without being the attribute or possession of a body, Foucault also suggests that "power is employed and exercised through a net-like organization", that "power circulates ... in the form of a chain", and that individuals "circulate between ... [the] threads [of power]; they are always in the position of simultaneously undergoing and exercising this power" (P/K 98).

For Faraday the omnipresence of force does not of course imply that force is evenly distributed over all points. Even though force is everywhere, Faraday allows for "clusters", "intensities" or "lines" of force, i.e. concentrations of force in various parts of the field. This idea can also be expressed by saying that the density of lines of force is different for different points in the field.[18] In an analogous way, Foucault proposes that we should not take power to be "the best distributed thing in the world, although in some sense that is indeed so" (P/K 99). While power relations permeate the whole social body of society, power relations can be more dense in some regions and less dense in others. Foucault comes close to Faraday's lines of force even terminologically when speaking of control mechanisms as "lines of penetrations" (*HS I* 42).

While these parallels suggest some ways in which the discourse of field theories of physics might provide heuristically useful concepts for the study of power along Foucauldian lines, they do only limited work in defending Foucault's omnipresence thesis against standard criticism. All they achieve in this respect is to show that omnipresence does not imply equal density, and that omnipresence need not preclude the possibility of resistance.[19] To see this, we only need to remember that Faraday's theory does not include the assumption that all vectors of force have to point in the same direction.

I shall not risk pushing the parallel too far, however. Instead, I shall make some topical remarks on the thesis of power as an omnipresent network. More precisely I shall (1) comment on the question of whether this amounts to a mystification of power, (2) take up the issue of whether this thesis is trivial, and (3) discuss the concepts linked to this thesis, i.e. the notions of chain and network.

(1) The first question that deserves attention is whether talk of power as an omnipresent network does not render the category of power useless as an analytical tool in the social sciences. Various writers – long before Foucault formulated his own views – have voiced the suspicion that the notion of power is of little use in our attempts to come to grips with social and political life. These authors claim that explaining social phenomena in terms of power "light[s] up very little of everything ... we see a little of the side, but the face remains obscure" (Minogue 1959: 283). Things seem to get even worse when the concept of power is enlarged so as to have as its extension the whole of society. It has been suggested that in this all-embracingness, the notion of power is no longer

opposed to anything, that when taken in this way power, in fact, collapses into the practice of the domain it was meant to illuminate (Turner 1989: 533–34). What remains after this radical expansion is a concept that explains everything and nothing, in other words, a concept with a mystical ring.

Admittedly, many more programmatic statements of Foucault – when taken out of their context of historical studies – have something of a mystifying ring. For instance, Foucault speaks as if power itself was an actor ("Power can retreat here ... invest itself elsewhere ..." (P/K 56)), says that "power is never localized here and there" and writes that power has no origin (P/K 199). Actually it is precisely the latter claim that leads him to reject the idea of developing any "theory" of power: "If one tries to erect a theory of power one will always be obliged to view it as emerging at a given place and time and hence to deduce it ..." (*ibid.*)

Yet despite the implausibility of the latter position, it seems to me that Foucault cannot be charged with collapsing the notion of power into the notion of society. Even though ultimately only the whole of this study can provide the evidence needed, be it said, nevertheless, already here, that Foucault's detailed definition of power, his distinguishing between relations of power, on the one hand, and relations of knowledge, exchange, communication, production and kinship, on the other hand, his more or less detailed descriptions of different forms, technologies, origins, structures and networks of power mechanisms, and finally his attention to the opposition between vicious and non-vicious forms of power, set him worlds apart from mystifications of power. After all, Foucault does provide us with a conceptual framework that can be – and has been – applied to detailed historical analyses.

(2) The Foucauldian version of the idea of power relations as penetrating all through the social sphere might, from a different perspective, be accused of triviality rather than mystification. After all, everyone knows from their own experience that power relations can be identified all around and between almost any interacting individuals.

Now, evaluating a thesis as to its degree of triviality is possible only by placing it against the larger backgrounds first of the historical moment when it was advanced, and second of the overall theory to which it belongs. Applying this to Foucault's omnipresence thesis, one cannot but conclude that in the French context of the 1970's, a time when structuralist Marxism loomed large, it was not trivial to advance the notion of the omnipresence of power. In this period, the very term 'power' had been reserved for the abstract level of class relations and structures, and the Althusserian Nicos Poulantzas distinguished even conceptually between "pouvoir" as the higher-level and genuine form of power of one class over another, and "puissance" as the uninteresting and politically irrelevant omnipresent relations of power between individuals (Poulantzas 1978a: 106). To state the omnipresence of *pouvoir* in this context meant proposing that higher-level and lower-level powers are interrelated enough to be referred to by the same concept. It was making a point, not stating the obvious.[20]

(3) Finally, we need to examine Foucault's way of formulating the thesis of the omnipresence of power in terms of chains and networks.

The fact that Foucault speaks of power relations as forming chains or nets should not come as a surprise given his emphasis on *relations* rather than *attributes* in the study and conceptualization of power. After all, a network is but a set of relations holding between a set of points, objects or individuals. A *Foucauldian power network*, however, has some specific features that are worth mentioning. Since Foucault regards power relations as internal-essential it is obvious that the individuals over which the network is defined cannot be thought of as existing apart from the network. In other words, the structural position of an individual, i.e. the sum of her (more basic) power relations to others, as well as the set of positions she can take up within the network are (at least in part) constitutive of the individual's identity.

It seems that a number of Foucault's suggestions concerning power and knowledge can perhaps be made more transparent by formulating them explicitly in the idiom of sociological network analysis.[21] For instance, recall that Foucault speaks of power relations as being interrelated, as being transformable into further new power relations by "matrices of transformations", and that he allows for power relations to "cross, .. sometimes cancel one another out, sometimes reinforce one another". Translated into the idiom of network theory which represents networks as directed graphs with nodes tied to one another by labelled arrows, these ideas amount to saying that a given network of power ties can be enlarged, reduced or transformed, where the latter possibility implies the introduction of new *types* of (and thus new labels of) power ties (or arrows).

Network concepts can perhaps also highlight Foucault's overall interests as well as illuminate his notoriously obscure statement that power "is the name that one attributes to a complex strategical situation in a particular society" (HS I 93). According to this suggestion, to describe power in a given society is not only to depict the different types of networks of power relations and mechanisms existing in the society, but also to portray the (types of) interrelation between various smaller networks, to study the stability and instability of networks as well as the causes for the stability or instability in question, and to sketch the ways in which what might be called "the identity tie" (i.e. the power/control of an individual over herself) is determined by its position in, or changes in, the network. A Foucauldian study of these networks will, first, label the ties according to the mechanisms involved, second, delineate how the introduction of new types of ties is possible, and third, show how the introduction of new types of ties changes already existing networks. Furthermore, a Foucauldian study will also investigate the relations between networks of social power and networks of knowledge, study whether and how these networks are interlocked, mutually supportive or isomorphic. Finally it will also address the question as to which macro-structures are latent and implicit in the myriads of microlevel relations. Thus genealogy performs a "reduction" of the multitude of individuals and relations to equivalent classes; that is to say, it distributes individuals into classes according to whether these individuals occupy an equivalent or similar position in the network(s), and it distributes relation tokens into classes according to their similarity and equivalence. In engaging in this operation, this investigation of course performs an operation that is also carried out by individuals and groups in society itself.

To be sure, this kind of investigation will not be able to answer in any absolute sense the question as to where power originates from, since its starting point will always be already existing networks of power. But it will address the question as to how these networks change, how they condition one another and how new types of networks emerge from earlier ones.

NETWORKS OF POWER AND INVISIBLE HANDS

In order to further appreciate the fact that Foucault's own advocacy of the omnipresence of power thesis leads to fruitful and interesting implications, we need to turn to his fourth thesis with respect to power: *Microlevel power relations and mechanisms are different, and thus to be distinguished, from macrolevel power relations between classes, on the one hand, and macrolevel power mechanisms, i.e. (systems of) coercive institutions, on the other hand. The emergence, persistence and effects of macrolevel relations and mechanisms are often unintended and unforeseen.*

Two distinctions appear in the first sentence of this thesis, the distinction between a macro- and a microlevel, and the distinction between relations and mechanisms. Neither of these distinctions have been fully attended to in the literature, and regrettably Foucault does not bring out clearly their importance to his enterprise.

Foucault speaks of "infinitesimal mechanisms" of power that become "invested, colonized, utilized, involuted, transformed, displaced, extended, etc. by ever more general mechanisms and by forces of global domination" (P/K 99). He also refers to these mechanisms as "micromechanisms" of power. "The more general mechanisms" of the last quotation are but the coercive institutions of the prison, the asylum, the clinic and the boarding school, institutions whose historical emergence and development Foucault investigates in *Madness and Civilization* and *Discipline and Punish*. Foucault calls these institutions and the systems they form "blocks", "apparatuses" or "dispositifs" (P/K 197). The last, not straightforwardly translatable word has several meanings, for instance, "an ensemble of parts that constitute an apparatus or machine", "an ensemble of measures that constitute an organization or plan", and "an ensemble of (material) precautions and means for carrying out a strategic, military operation"[22]. The last-mentioned meaning harmonizes especially well with Foucault's liking for military terminology.

Power mechanisms do not have a life of their own. Rather they are employed within micro- or macrorelations of power. The latter distinction is hinted at in the following methodological advice:

> We need to identify the agents responsible [...] those which constituted the immediate social *entourage*, (the family, parents, doctors, etc.) and not be content to lump them under the formula of a generalized bourgeoisie (P/K 101).

To be sure, Foucault does not rule out the possibility of speaking of the bourgeoisie or the proletariat as classes: he writes, for instance, that "the bourgeoisie was interested in power, not in madness" (P/K 102). However, he holds that the attention given to the macrorelations between classes must not lead us to disregard the microrelations with respect to which macrorelations are but abstractions or reductions. Thus the opposition Foucault focusses on is one between microlevel power relations between singular agents, or small groups, and macrolevel power relations between classes.

The distinction between the micro- and the macrolevel also surfaces in Foucault's writings as the opposition between "tactics" and "strategies". For Clausewitz tactics was "the theory of how to use forces in a battle", and strategy "the theory of how to use battles for the purposes of war" (1918: 69). In the same vain, Foucault refers to tactics as the local employment of "infinitesimal mechanisms" in exercises of power between individuals, and to strategy as the combining, using, and influencing of these local relations of power for the purposes of social war. The interdependence of these two levels is suggested in the following passage:

> ... the great strategies of power encrust themselves and depend for their conditions of exercise on the level of the micro-relations of power. But there are always also movements in the opposite direction, whereby strategies which co-ordinate relations of power produce new effects and advance into hitherto unaffected domains (P/K 199–200).

Before going further it might be helpful to summarize the distinctions just outlined in the form of a table (Figure 13).

	relations between	mechanisms
macrolevel (strategy)	(1) classes	(2) *dispositifs*
microlevel (tactics)	(3) individuals, social *entourage*	(4) infinitesimal

Figure 13

Foucault's central interest is with the birth and growth of (2). Thus the question is: how is this birth and growth to be explained? Foucault reproaches Marxism for trying to explain too much in terms of relations between classes or the mode of production. For instance, he opposes attempts to explain the prison or the asylum in terms of bourgeois interests. The bourgeoisie and its objectives, or the emergence of coercive institutions, cannot be defined and understood without engaging in an "ascending analysis" from the microlevel up (P/K 99). Put differently, to reduce the multiplicity and variety of power networks to the opposition between two social classes is to bracket too many of precisely those (types of) power ties that are needed to account for the emergence and persistance of coercive institutions.

In line with this criticism Foucault's own historical analyses do not seek to explain (2) in terms of (1), but rather to explain (2) via (3) and (4), and to explain (3) in terms of (2) and (4). It would seem that Foucault suggests a twofold explanation for the existence of *dispositifs*, although it is hard to show that he has ever in fact presented such an explanatory account in any detail. What a historical account of the existence of a *dispositif* calls for is, first, a description of the *structure*, *profile* or *composition* of the coercive institution, and second, an explanation of the *conditions of its possibility*.

To tackle the first point is to move from (4) to (2). It is – as already mentioned – to study how "infinitesimal mechanisms" of power are, or have been, "invested, colonized, utilized, involuted, transformed, displaced, extended, etc." by and within systems of *dispositifs*. One investigates *dispositifs* both as social laboratories, and as the results of this laboratory work: *dispositifs* are spaces where experimentation with small, closed power networks, or social "microworlds" is possible, spaces where new technologies of more efficient, more extensive, more diverse forms of control, manipulation, authority, coercion, punishment and deterrence are developed.

Suffice it here to mention only Foucault's favorite example, "panopticism" as a "technological invention in the order of power, comparable with the steam engine in the order of production" (P/K 112). This example can also serve as an illustration of the more basic nature of Foucauldian power mechanisms as compared with the traditional "species" of power. As is easy to appreciate, the panopticon can be used to strengthen or create relations of coercion, authority, etc. It can also be used, say, to manipulate and coerce, both at the same time.

Developed and applied in prisons, hospitals, armies and factories, the panopticon is a type of power network in which one individual, or a few, can keep a much larger group under constant surveillance. Members of the latter are isolated from one another yet are at all times visible to the controller. (For instance, in the prison, cells are arranged in a circle around a central tower, the cell side facing the tower having bars rather than a wall.) The effectiveness of this technology – whose basic structure can be adopted and modified for the needs and resources of different environments – does not simply lie in the extent of visibility it provides or in the small number of individuals needed for control. Its most intriguing feature is the effect the constant surveillance has upon the controlled: they can be expected to internalize the controlling gaze and thus to control themselves. And what is more, this technology is truly 'closed' in that even the individual at the controlling end is herself controlled, too: if she neglects her duty and the network breaks down, she will be the first to suffer for it. As Foucault puts it:

> This indeed is the diabolical aspect of the idea and all the applications of it. One doesn't have here a power which is wholly in the

hand of one person who can exercise it alone and totally over the others. It's a machine in which everyone is caught, those who exercise power just as much as those over whom it is exercised (P/K 156).

To address the second issue, the *conditions of the possibility* of a coercive institution, is again a double task. First, it is to identify "the immediate social *entourage*", that is, the (groups of) agents who conceive of a given *dispositif* as enhancing the scope, intensity, amount and efficiency of their own exercise of power over others. In other words, these actors take the *dispositif* in question as changing prevailing networks of power in their favor, or as upholding an existing network of power to their advantage, e.g. by providing these individuals with new, more efficient mechanisms of power. (Step from (4) to (2).) Perhaps Foucault's most convincing attempt regarding this task is his editing – with Arlette Farge – of a selection of the famous *Lettres de cachet* (1982), letters that in the seventeenth century could be sent directly to the King asking for the imprisonment of one's family members, neighbors, or employees. As the documents bring out impressively, people from all social strata eagerly seized upon this possibility to enhance their authority within their family, their neighborhood and their workplace. Thus already existing power relations led people to "willingly invest themselves"[23] into this new technology of control and authority.

Second, to explain the conditions of the possibility of a coercive institution is also to explain how individuals or groups become constituted as either inmates and victims, or as bottom level officials. (Here we thus move from (2) and (4) to (3).) This question is of course natural to ask given Foucault's thesis of power relations as internal-essential. In his historical studies Foucault is more concerned with the inmates than with officials, even though he mentions that the problem of training officers, teachers and foremen became an acute one by the eighteenth century and that the State long lagged behind the Church in acquiring the lower cadres needed for control and supervision (P/K 157). As concerns the issue of inmates, Foucault suggests, for instance, that the prison functions in a self-defeating – or self-fulfilling – way: the networks of power mechanisms within this *dispositif* constitute precisely those kinds of personalities and identities that the prison is allegedly meant to reshape.

To complete the discussion of the interrelations between relations and mechanisms, between the macro- and the microlevel, let us note that Foucault is aware of the possibility that the emergence of (2) can be unforeseen on the level of (3), i.e. unforeseen by the individual actors involved.[24] Foucault indicates his position in the following passage:

> ... the rationality of power is characterized by tactics that are often quite explicit at the restricted level where they are inscribed (the local cynicism of power), tactics which, becoming connected to one another, attracting and propagating one another, but finding their base of support and their condition elsewhere, end up forming comprehensive systems: the logic is perfectly clear, the aim decipherable, and yet it is often the case that no one is there to have invented them ... an implicit characteristic of the great anonymous, almost unspoken strategies ... (HS I 95).

Even though one might want to question whether the use of the notion of "strategy" is a happy terminological choice here, the point itself should not come as a surprise. After all I have already mentioned Foucault's claim that the power networks within society change constantly, and that the introduction of new elements at one point in the network lead to re-arrangements throughout the whole network.

Unfortunately, over and above occasional hints like the quoted passage, Foucault has not provided us with a theoretical account of invisible hand or counterfinality explanations. Neither do his historical studies include unambiguous cases of such explanations from which one could infer his conception. This lacuna is unfortunate, particularly since Foucault's vision of modern society as "panoptical" – i.e. as a society where everyone is controlled by someone, without there being someone who controls everyone – obviously stands in need of an invisible hand account.[25] It seems clear that any rational utility maximizer (or group thereof) who can design such a society could also design a society where only he, she or they is or are in command.

As Foucault's emphasis upon "ascending analysis" already suggests, we may surmise that had Foucault developed his conception more fully, he would have come to speak out in favor of *aggregate* rather than *functional mold* invisible hand explanations. Whereas aggregate mold explanations take intentions of actors as their input in order to yield, via an assumed complex process, an unforeseen social phenomenon as their output, the functional mold version explains the *raison d'être* of a social phenomenon in terms of its function within a social system.[26] Further evidence for this suggestion can be marshalled from an interview where Foucault sides with methodological individualism by assenting to the claim that "there are only ever transitory coalitions, some of which immediately break up, ... others of which persist, but ... strictly speaking individuals ... [are] the first and last components" (P/K 208). Furthermore, in the same interview, Foucault fiercly denies wishing to model his explanations of *dispositifs* and their effects upon biological concepts like organism; thus he rejects precisely the notions upon which *functional mold* invisible hand explanations are modelled:

> ... I have never used the metaphor of the organism. ... When I speak of strategy, I am taking the term seriously: in order for a certain relation of forces not only to maintain itself, but to accentuate, stabilize and broaden itself, a certain kind of manoeuvre is necessary. The psychiatrist had to manoeuvre in order to make himself recognised as part of the public hygiene system. This isn't an organism, any more than in the case of the magistrature, and I can't see how what I'm saying can imply that these are organisms (P/K 206).

11. THE GENEALOGICAL CONCEPTION OF POWER II: SOCIAL POWER AND SCIENTIFIC KNOWLEDGE

The Productivity of Power

Little has been said so far about the two topics that figure most centrally in Foucault's genealogy, to wit, the productivity of power, and the relation between power and knowledge.

Foucault never tires of reminding us that not all exercises of power have merely the effect of restraining, limiting and restricting. In his words, the effects of power are not in all, and perhaps not even in most, cases, "negative". Rather, exercises of power, power mechanisms and power networks, can also have productive or "positive" effects such as the power subject's acquiring new skills, her adopting new beliefs, or her being instrumental as an object of experimentation (and thus her being involved in the growth of knowledge).

While being one of Foucault's most often cited genealogical ideas, the thesis of the productivity of power, at least in its general form, is neither original nor surprising. After all, manipulation has always been conceptualized as the *production* of wishes or beliefs in the power subject. However, even though the idea of the productivity of power is uninformative in its general form, it gains in substance, when related to Foucault's view of the relation between scientific knowledge and social power. If it can be argued that the productivity of (social) power is somehow internally related to the growth of scientific knowledge, we obtain a thesis which is non-trivial. In this section I shall argue that such conception naturally emerges on the basis of Foucault's suggestions.

Thus, as concerns the relation between social power and scientific knowledge, the thesis that seems natural to attribute to Foucault can be formulated as follows: *The relation between scientific knowledge and social power is internal-essential rather than external-accidental.*

What makes it inviting to attribute this position to Foucault are his statements to the effect that although social power and scientific knowledge cannot be separated, their relation is nevertheless neither a conceptual one, nor one of identity; in other words, the relation between social power and scientific knowledge is, at least in part, causal:

> 'Truth' is to be understood as a system of ordered procedures for the production, regulation, distribution, circulation and operation of statements. 'Truth' is linked in a circular relation with systems of power which produce and sustain it, and to effects of power which it induces and which extend it (P/K 133).

That Foucault does not conceive of the power/knowledge relation as being merely accidental is obvious from statements like "truth is already power" (P/K 133), or "there is no point in dreaming of a time when knowledge will cease to depend on power" (P/K 52). Their identity is, however, rejected in the following passage:

> ... when I read – and I know it has been attributed to me – the thesis, "Knowledge is power", or "Power is knowledge", I begin to laugh since studying their *relation* is precisely my problem. ... The very fact that I pose the question of their relation proves clearly that I do not *identify* them (PPHC 43).

In order to work our way towards a defense of an internal-essential relation between social power and scientific knowledge, let us start from observing that all forms of successful power exercises presuppose or involve knowledge and belief. For instance, in cases of coercion the power exerciser is well advised to know – at least roughly – the value of the booty to his victim; in cases of manipulation the coercer needs to know something about human psychology; in cases of authority the power subject relies on the allegedly superior knowledge of the authority; and in cases of deterrence one agent wants the other to know what will happen once the latter either starts or changes a specific course of action.

However, when asking for the relation between knowledge and power, we are obviously after something more, or different, from these last-mentioned relations. What we are interested in, and what seems to be the contested question, is not the occasional employment of knowledge in exercises of coercion, nor – to mention another relation between knowledge and power – the connections between knowledge and power-*to*, i.e. power in the broad sense of capacity to bring about some state of affairs. What we are interested in is rather the role that social power plays in the shaping of scientific knowledge, the problem as to whether *all* scientific knowledge is related to power, and the question of whether science would be (or is) possible without the mechanisms and effects of power exercises.

In order to bring out more clearly what is unsatisfactory about approaches to the problem of power/knowledge that do not meet these conditions, let us look briefly at one more recent attempt, by Barry Barnes, to argue for an identity, or near-to-identity, between power and knowledge. Pointing out the weakness of this proposal can pave the way towards what I submit are more compelling perspectives or arguments: the scientific community as a field of agonistic struggles, the interdependence between the emergence of *dispositifs* and some of the human sciences, science as constituted by exclusion, and social power as a determining factor of forms of scientific knowledge. I shall conclude by asking how the internal relation between power and knowledge is to be understood in modal terms, and what its implications are for the issue of realism *versus* anti-realism.

BARNES' PROPOSAL

The line of reasoning of Barnes's *The Nature of Power* (1988) takes its starting point from the "common sense concept of power". According to this concept, power is "the capacity to do work", or, more precisely, since one and the same power usually can achieve more than one goal, "the *generalized* capacity" to bring about (or hinder the coming about of) states of affairs (*ibid.*, 2–3).

To see how Barnes uses this general concept of power as capacity in his definition of *social* power, we need to introduce two more of his suggestions. On the one hand, Barnes argues that we can look upon a society *either* as a set of routines and actions, *or* as a distribution of knowledge: "... one characterization is as good as the other" since "human beings act calculatively on the basis of what they know" (*ibid.*, 46). On the other hand, Barnes proposes that the power of a collective is larger than the power of all its individuals taken singularly because the capacity to act increases through coordination and exchange. Social power is then the added capacity for action accruing from coordination, i.e. from forming a society. Thus, since the latter is *eo ipso* a distribution of knowledge, Barnes can write that ...

> social power is the added capacity for action that accrues to individuals through their constituting a distribution of knowledge and thereby a society (*ibid.*, 57).

However, this definition only tells us what social power *is*. From this definition of "the nature" of social power, Barnes distinguishes the issue of its possession. According to *The Nature of Power*, "social power is *possessed* by those with discretion in the direction of social action, and hence predominantly by those with discretion in the use of routines" (*ibid.*, 58). Furthermore, Barnes stresses that to be and to be known to be socially powerful are one and the same thing. This is so, since, for Barnes, society is not just any distribution of knowledge, but rather a "self-referring" and "self-validating" one (*ibid.*, 46). In other words, the knowledge in question is knowledge about the roles of various individuals in that society, knowledge about prevailing routines, norms and practices in society. Moreover, this knowledge is true precisely because it is constitutive of society itself:

> To come to believe something about the status of an individual is to do two things at once: it is to accept a *claim about* his status and at the same time to contribute to the *constitution* of his status (*ibid.*, 49).

For instance, to know that the chairperson in your department is powerful is to participate in making or keeping him powerful.

From these foundations Barnes advances three theses. According to the first, (social) power is omnipresent. Even though social power is possessed only by a minority, social power "lies outside" the power holder, "right down there among the supposedly powerless". It is "only" (!!) the discretion over this power that is concentrated (*ibid.*, 62-63). Capacities to act are everywhere; it is merely the discretion over them that is unequally distributed.

Second, knowledge and social power "are the same things under different forms of awareness, and a change in the one does not cause a change in the other but rather entails it" (*ibid.*, 169–70). Since the given distribution of knowledge constitutes the powerful and provides them with the social power qua discretion over the added capacities of their polity, knowledge and social power are but two sides of the same coin. To be socially powerful is to be a certain kind of referent of some given distributed knowledge. In other words, to be socially powerful is to be known to have, and thus have in fact, discretion over added capacities accruing from the coordination of actions, that is, accruing from the coordination or specific distribution of knowledge.

Third, and finally, the identity of (social) power and knowledge does not imply a "simple correspondence" between specific singular "bits" of knowledge and singular "bits" of social power. Rather:

> The existence of a specific power, available at a specific point in the social order, may be a matter of the existence of innumerable items of knowledge, spread over every part of the social order. The existence of a specific item of knowledge, at a specific point in the social order, may be implicated in the constitution of numerous powers available at many points (*ibid.*, 169).

Barnes's ideas are instructive for our concerns in so far as his book illustrates especially clearly that a one-sided attention to power-*to* rather than power-*over* does only very limited work in helping us to understand the relations between scientific knowledge and social power as domination.

Note, first of all, that Barnes indeed bases his argument crucially on the revisionary redefinition of power as capacity. Phenomena like coercion, authority, control, etc. thus remain inevitably bracketed in his account. To be sure, in Barnes's case, power-*over* is preserved under the new conceptualization of "discretion over social power". Yet it is not difficult to point out that this suggestion is counterintuitive: not only does this new conceptualization force Barnes to write that it is "only" discretion rather than power which is hierarchized, but this conceptualization also leads straightforwardly to his claim that every change in (the distribution of) knowledge *entails* a change in (the distribution of) power. Even if we grant him this claim for the case of capacity – and even there it does not seem free of implausibility – it certainly does not need to hold for discretion over social power(s) as added capacities. Social power as domination (i.e. Barnes's discretion) is certainly more stable than this suggestion implies.

Secondly, even though Barnes explicitly includes scientific knowledge of both the natural and the social sciences within his notion of society-shaping knowledge, and even though he regards his investigations as closely linked to the sociology of (scientific) knowledge (*ibid.*, 170–71), his revisionary account of power does not even provide hints as to how we should relate social power (in its standard sense as power-*over* others) to the production and distribution of knowledge. All we can say on the basis of Barnes's book on the relation between scientific knowledge and power, is that scientific knowledge is part of the distributed knowledge of society, that it is in part constitutive of that society, and that it is as such internally related, nay, identical, with capacities. But these claims are either trivial - if one accepts Barnes's revisionary definitions - or false - if one does not accept his leaving the concept of power-*over* to the role of a 'poor relative'.

DEFENDING "POWER/KNOWLEDGE"

Turning from Barnes to Foucault, a first way to approach the issue of science and power in a Foucauldian spirit is to place it within the context of his overall theory of power. As we saw above, Foucault advances the theses of the omnipresence, and of an ascending analysis of power. Since science, or rather scientific communities, are part of the social field, obviously both ideas also apply in its or their case. Unfortunately, Foucault has not paid much attention to this application. He merely informs us of his siding with Nietzsche in believing that the scientific field is a field of struggles, and more precisely, that the "devotion to truth" results from "the passion of scholars, their reciprocal hatred, their fanatical and unending discussions and their spirit of competition" (NGH 78).

Yet while Foucault himself is disappointingly brief on this issue, it seems easy enough to suggest questions for the social-historical study of science that are but a transferring of Foucault's general ideas on power to this special case. Thus taking our lead from the notion of power networks and the distinctions between relations and mechanisms, the micro- and the macrolevel, we might ask how the scientific community as a power network relates to the scientific community as a knowledge network (a network for the production and circulation of facts and theories); what are the infinitesimal mechanisms of power in the scientific community; what was the historical process by which they came to be organized, combined and structured in institutions like the school, the university and the laboratory; how was this organizing process possible, i.e. which - partly cynical - interests of which groups made these organizations seem attractive; which - partly cynical - interests of which groups and agents uphold and re-enforce these organizations today; and how are individuals disciplined by power relations in this field for these individuals to find these relations and mechanisms useful or inevitable.

To be sure, not all of these questions are new. Several of them have already received at least partial answers in some more recent studies in the philosophy and sociology of science. These enquiries study the use of humans and animals as experimental 'material', the agonistic struggle among scientists for recognition, credit or credibility, identify the disciplinary techniques employed, the mechanisms by which scientists monopolize important resources like new instruments, mechanisms that keep younger researchers obedient and loyal, and mechanisms by means of which one prevents competitors receiving funds or gaining access to important publication channels. However, while making for fascinating reading, it can hardly be claimed that these investigations are more than just a beginning. Thus, although a start has been made to increase our understanding of power in the laboratory, not much work has yet been done, for instance, on the mechanisms of power in university departments: the sociological anthropologists are not yet at large among philosophy tribesmen. The perhaps best-known study that takes some first steps in this direction, Pierre Bourdieu's masterly *Homo Academicus*, does not direct itself to the microsociological level even though it makes many interesting observations relating to it. For example, Bourdieu is a perceptive observer of the ways in which French professors can keep their pupils in a relation of dependence by determining their career rhythm, their chances to publish, the time of their doctorate, and their appointment to short-term positions. The character or attitude that these mechanisms produce is, in Bourdieu's happy phrase, "the docile and submissive, even somewhat infantile, attitude which characterizes the good student of all eras" (Bourdieu 1988: 88).

However, treating science as an agonistic field of struggle among scientists does not yet in itself establish that scientific knowledge *qua* product of the scientific activity is internally-essentially interlocked with social power. The defender of the received view might still hold on to his or her position by distinguishing between the social context of science and scientific knowledge itself. In order to undermine his or her position more radically, we therefore need to turn to three other arguments.[27]

A disquieting suggestion

The first argument is little more than a disquieting suggestion that applies perhaps only to some of the human sciences. Scientific knowledge of the human sciences might be internally related to power mechanisms because the very subject matter of those sciences is, at least partly, constituted by mechanisms of power that those human sciences help to justify, use as instruments, and have taken over from coercive institutions. Put differently, scientific knowledge of the human sciences might just be an enormous self-fulfilling prophecy because the mechanisms upon which it is based are but versions of those mechanisms of control, surveillance and manipulation that have constituted its objects.

This disquieting suggestion arises from Foucault's study of the history of psychiatry and criminology. Foucault argues that the symptoms of mental disease or social deviance are diagnosed and treated by the same mechanisms of surveillance, control, drill and categorization that figure centrally in the mental clinic or the prison, on the one hand, and that were created and refined in and for these coercive institutions by psychiatrists and criminologists, on the other hand.

Science and exclusion

Another argument for an internal link between social power and scientific knowledge is one that looms large in all of Foucault's genealogical writings: scientific discourses constitute themselves only by exclusion and prescription, i.e. by drawing a line between scientific and non-scientific knowledge. Foucault speaks of the excluded knowledge as "subjugated knowledge", as "the historical contents that have been buried and disguised in a functionalist coherence of formal systematization" and that ...

> have been disqualified as inadequate to their task or insufficiently elaborated: naive knowledge ... beneath the required level of cognition or the scientific ... [e.g.] that of the psychiatric patient, of the ill person, of the nurse, of the doctor ... of the delinquent ... (P/K 81).

Foucault also confronts those who seek to establish Marxism as a science with the question, "what types of knowledge do you want to disqualify ...?" (P/K 85).

As Foucault sees it, exclusion functions through a discursive "order", a "regime of truth" (P/K 133). Concerning the latter, Foucault tells us that it is a "circular relation" between truth "as a system of ordered procedures for the production, regulation, distribution, circulation and operation of statements", on the one hand, and "systems of power which produce and sustain it [i.e. truth] ... and effects of power which it induces and which extend it" (*ibid.*), on the other hand. Put more succinctly, "one is within the [region of] the truth only when being obedient to a discursive police" (LD 37).

In his *L'Ordre du Discours*, his inaugural lecture at the Collège de France in 1970, Foucault suggests a systematization of the different forms of exclusions and prescriptions by which a field of science is constituted. A first main category are "procedures of *exclusion*" (LD 11). Within this category Foucault distinguishes further between "interdictions" concerning certain objects, styles of arguing, and individuals or groups, a "rejection" of what is defined as irrational and mad, and a "partition" between the true and the false – a partition that amounts to an exclusion of what does not conform to some specific conception and standard of truth (LD 11–15).

A second main category, "internal procedures", relates to prescriptions on how statements have to be formed and formulated in order for them to qualify as scientific within a given field of study. Such prescriptions bear upon the manners according to which one is to produce "commentaries" on the work of other scientists; the ways one is expected to group statements and theories around the originators or "authors"; and the list of objects or processes to be talked about, or conceptual and technical instruments to be employed. Foucault refers to this last item as "discipline" (LD 23–31).

Finally, the third main category refers to various procedures that further define and regulate competent membership, that is, "conditions of admission". These are, first, the "ritual", i.e. the criteria of proper social/discursive behavior; second, the "discursive societies", i.e. visible or invisible closed colleges all of which put specific demands upon their members; third, the "doctrine", i.e. the – at least verbal – adoption of, or abiding by, specific religious, political and philosophical views, and fourth and finally, "education" (LD 38–46).

That this categorization is a useful one, can easily be vindicated by relating it to work done in the sociology or history of science, fields that tend to be stronger in collecting empirical materials than in providing theoretical frameworks. Steven Shapin's and Simon Schaffer's study of Boyle's introduction of experimental science as a new language-game, *Leviathan and the Air-Pump. Hobbes, Boyle, and the Experimental Life* (1985), is a case in point. Shapin and Schaffer distinguish only vaguely between "material", "literary" and "social technologies" employed by Boyle (*ibid.*, 25). Using Foucault's distinctions seems to give a more fine-grained picture, however.

As concerns *exclusion*, Boyle's new science was defined by *interdictions* with respect to *objects* like insubstantial beings and spirits, self-moving matter, and metaphysics; interdictions of *styles of arguing* like deductivism or *ad hominem* arguments typical of philosophical treatises at the time (Hobbes); and interdictions of *groups or individuals* like alchemists, rationalist philosophers and hermeticists. *Partition* between the truth and the false took the specific form of accepting only the probable as true, and of rejecting all claims to absolute truth as *per se* unscientific (*ibid.*, 58, 61–69, 72, 167, 176, 203).

Internal procedures consisted of specific demands with respect to *commentary*, the latter being required to be modest, and abstaining from any sharp critical overtones; specific demands with respect to the category of the *author*: experiments must be attributed to individuals not the experimental community as a whole (i.e. only individuals can fail, not the scientific experimental community); and specific demands of the *discipline*: use of the air-pump as the central instrument of experimentation, and acceptance of the air-pump as air-tight for all practical purposes, rejection of all questions as illegitimate that cannot be handled experimentally, abiding by the distinction between experimental matters of fact and their ultimate physical causes or explanations, supporting a sharp opposition between actual and thought experiments, regarding the spring of air as the central object to be investigated, taking the reports of eye-witnesses as decisive, and the employing naturalistic representations in one's publications (*ibid.*, 29, 45, 51, 55, 56, 60–69).

Finally, *conditions of admission* included those of the *ritual*, i.e. a civil and liberal tone, an attitude of trust and modesty, a functional style of writing, the abstention from preconceived expectations as regards experimental outcomes, and the posing as a "priest of nature". Admission was also restricted by the Royal Society as a closed *discursive society*: eye-witnesses were restricted to lawyers, priests and hand-picked individuals, peasants were regarded as unqualified, testimony of experimental outcomes by non-members were rejected, and only *educated*, or disciplined witnesses were acceptable. Finally, *doctrines* to be adopted included the rejection of both vacuism and plenism, the belief in the reliability of sense-perception as a theory-neutral foundation of science, fallibilism, a mechanistic conception of nature, and a religious conception of God as non-corporeal (*ibid.*, 36–42, 58–76, 171, 206, 218, 319, 336).

The fact that the interesting observations of *Leviathan and the Air-Pump* can be reconstructed and put more succinctly in Foucauldian terms, also suggests a reformulation of the book's main topical thesis. In their rather brief dealings with the theoretical implications of their study, Shapin and Schaffer conclude from their detailed account of Boyle's program and its relation to the wider political situation of the time, that science itself is to be analyzed in terms of politics. The authors hold that in order to understand the possibility of the creation of scientific knowledge we have to understand the working of scientific communities and their structures. To deal with the latter, however, is to study the interplay of three factors: "(1) the polity of the intellectual community; (2) the solution to the practical problem of making and justifying knowledge; and (3) the polity of the wider society" (*ibid.*, 342). These factors are interconnected in that making and justifying knowledge is possible only against the background, or within, an intellectual community with specific rules of exclusion and prescription. Furthermore, scientific knowledge usually spreads out from the scientific community into the wider polity of society where it is used as a resource. And finally, the introduction and maintaining of any specific scientific forms of life is dependent on whether its advocates succeed in mobilizing support for it in the wider society (*ibid.*, 342).

One need not disagree with this conclusion to suggest that it is perhaps brought into sharper focus by being conceptualized in terms of social power rather than in terms of "politics". Relying on the notion of power – regardless of whether in a Foucauldian or some other version – at least immediately prompts us to pose a limited list of precise questions. These questions concern topics like responsibility, invisible hand effects, networks, forms of sanctions, scopes, amounts, extensions, and mechanisms of power, means, resources, distribution and intensity of power, as well as the constitution of subjects in and by power mechanisms.

Shifting from a vague notion of politics to a somewhat more elaborated notion of social power, we can reformulate Schaffer's and Shapin's program by saying that to study the possibility of knowledge is to investigate the relations between the social power network of society and the distribution of social power within and outside of science. The production and justification of knowledge is possible only against the background of, or within, an intellectual community as a network of social power. Scientific knowledge usually spreads out from this community into the wider society, changing, enhancing and modifying existing power relations by providing various individuals or groups with new or improved power mechanisms, resources and means. Finally, the introduction and maintaining of specific scientific forms of life – and the power structures characteristic of them – depend upon whether or not they succeed in convincing a sufficient number of (powerful) individuals and groups of their usefulness, where this usefulness in part again is measured by their production of, and improvement of power resources and means.

Reformulated in this way, it is immediately clear not only that "power is productive" but also that the relation between scientific knowledge and social power is internal-essential rather than external-accidental.

An argument from underdetermination

Even though the last argument seems sufficient to establish an essential link between social power and scientific knowledge, it can perhaps further be strengthened by another line of reasoning. This line of reasoning is implicit in the argument above – and actually implied in part by Shapin and Schaffer's book. This third argument proceeds from the existence of underdetermination in science to, first, the inevitable involvement of social power in scientific decision making, and to, second, the idea that social power shapes our forms of knowledge. One might perhaps say that this argument is hinted at by Foucault when he stresses the contingent character of scientific and social development, on the one hand, and the pressure of power on science, on the other hand:

> ... things 'weren't as necessary as all that'; it wasn't as a matter of course that mad people came to be regarded as mentally ill; it wasn't self-evident that the only thing to be done with a criminal was to lock him up; it wasn't self-evident that the causes of illness were to be sought through the individual examination of bodies ... (QM 104).
>
> Power never ceases its interrogation, its inquisition, its registration of the truth ... (P/K 93).

However, the argument from underdetermination is best presented without trying to squeeze it out of Foucault's sketchy remarks.

The starting point, or first premiss, of the argument, i.e. the existence of underdetermination in science, more specifically, the underdetermination of theory by observation, and the underdetermination of the interpretation of experiments, the choice of theory and the choice of research programs by principles of rationality, is hardly controversial. (Even though one might still debate the *extent* to which underdetermination plays a role.)

The second premiss states that underdetermination in science is overcome by interests and that these interests are power interests. I regard this premiss as highly plausible based on the number of case studies that sociologists and historians have provided us with over the last decade or so. It appears that, when faced with choices, scientists tend to adopt interpretations, theories or lines of research that both enable them to monopolize areas of investigation, and that offer them either opportunities to dismiss the work of competitors as unfruitful, unscientific, and unworthy of funding, or opportunities to answer to the social demands of a particular class, the state, a political party or a church (Shapin 1982).

The third premiss is the hypothetical sentence that if there is, or were factors *F*, that determine what we regard as knowledge, good science and scientific exemplars or paradigms, then the relation between these *F*'s and knowledge is, or would be, internal-essential.[28]

Now given these three premisses, my conclusion should no longer come as a surprise. Since underdetermination is overcome by non-scientific interests, and since these non-scientific interests are power interests, therefore power is one of the factors *F* that are internally-essentially related to scientific knowledge. Since science *qua* social field is, and always has been, penetrated by power relations, and since the choice of theories and research programs has thus inevitably been strongly influenced by the interest to increase power – be it the power of individuals, classes or the military power of the state – then it becomes indeed implausible to think of social power and scientific knowledge as merely accidentally related.

Realism an Modality

To conclude this rather longish discussion of the internal-essential relation between power and knowledge, let us, at least briefly, ask, first what all of this amounts to for the issue of realism *vs.* anti-realism, and, second, how, in more precise modal terms, this internal-essential relation is to be understood.

I regard it as a distinct merit of the various arguments presented above that they do not prejudge anything concerning the tricky problems of realism or anti-realism in the philosophy of science.[29] What Foucault's position on this question might be is hard to decide on the basis of his writings – at least as far as the natural sciences are concerned. (I shall return to this topic below.) Be this as it may, both realist and anti-realist philosophers can accept the thesis of an internal-essential relation between scientific knowledge and social power. Nothing in the above arguments shows that there is no true knowledge, that scientific realism is wrong, or that in and through engaging in scientific work we do not find out things about parts of reality. To be sure, if these proposals are accepted, a study of social power cannot be excluded from the question of why we model reality in one way rather than another, of what pieces of reality we come to know, or of why science develops in one direction rather than another. But all this does not amount to saying that science is dealing – always and necessarily – with fictions.

I have already pointed out that an interesting and challenging argument for an essential relation between power and knowledge must not stipulate this relation by fiat, i.e. by providing revisionary definitions of the two *relata*, definitions on the basis of which the relation between social power and scientific knowledge becomes a conceptual one. This demand, however, calls for an answer to the question of what the internal-essential, but non-conceptual, relation between social power and scientific knowledge amounts to in modal terms. It should be clear, in any case, that since this relation is not a conceptual-logical or analytical one, it cannot hold in all possible worlds. After all, we can imagine a world of angels with scientific interests, or a world of humans in which all knowledge is obtained through divine revelations. Yet the fact that the relation between social power and scientific knowledge does not hold in all possible worlds does not oblige us to treat this relation as an accidental and contingent one. This is because we can preserve a notion of necessity here by restricting the possible worlds to *socially possible* ones. That is to say, the thesis of the internal-essential relation is tantamount to the truth of the following counterfactual:

(*) If the scientific communities were not networks of social power relations, and if they were not connected to various social power networks of the wider society, scientific knowledge as we know it would not exist.

This counterfactual (*) amounts to saying that in any possible state of affairs in which scientific communities are not social power networks, and which resemble our actual state of affairs as much as is permitted by the absence of social power relations in science, scientific knowledge as we know it does not exist.[30] Put differently, while not being a *logically* necessary relation, the relation between social power and scientific knowledge is *socially* necessary. In other words, the relation holds in all those possible worlds which are socially similar to our world: they contain humans or other intelligent creatures that resemble us in so far as they do not obtain knowledge by revelations, they are not angels or saints, they live in a world of scarce resources, and they face the problems of underdetermination of scientific theories by observation and standards of rationality.

12. GENEALOGICAL RESEARCH STRATEGIES

Introduction

Proceeding from the question of what power *is* or *how it works* to the question of *how it is to be (or can be) investigated*, it is clear that the latter issue cannot but be closely related to the former. It is hardly surprising, therefore, that in dealing with the Foucauldian theory of power we already encountered a whole number of proposals on how power might be studied. For instance, we saw Foucault attending to power mechanisms and exercises; employing models of war and physics; giving heed to mechanisms of ecological control by architectural measures; focussing on the training or disciplining of bodies; studying the ways power relations and mechanisms shape the personal identity of the coercer and the coerced; investigating power as a network of power relations; examining the interconnections between microlevel and macrolevel power relations; analyzing coercive institutions as social laboratories for the development of power mechanisms; taking one's starting point from individuals and groups rather than classes ("ascending analysis"); relying on invisible hand or counterfinality explanations; making use of concepts like "tactics" and "strategies"; considering the interrelations between social power and scientific knowledge; taking into account the involvement of interests; (re-)conceptualizing science as a social process of exclusion and regimentation ("orders of discourse"); and rehabilitating excluded, "subjugated" knowledge.

Many of these ideas are also alluded to in all of the four major places where Foucault explicitly spells out his methodological "demands", "rules" or "precautions".[31] Since all of these rules flow fairly straightforwardly from the theses discussed above as the Foucauldian theory of power, there is little need for going over them in detail. What calls for exploration and discussion, however, are some more general methodological ideas (and the reasoning behind them), ideas that motivate Foucault's choice of the label "genealogy" for his investigations into power and knowledge. In other words, what needs to be investigated are the central principles and suggestions that jointly constitute the Nietzschean background, or the Nietzschean dimension, of Foucault's enterprise.

Foucault himself describes the intention behind his genealogical work as seeing "[...] to the extent that it is possible, with the aid of Nietzsche's texts - but also with anti-Nietzschean theses (which are nevertheless Nietzschean!) - what can be done in this or that domain" (PPHC 251). He also suggests that "the genealogy of morals" might stand as a title over his own work, and he praises Nietzsche for being "the philosopher of power, a philosopher who managed to think of power without having to confine himself within a political theory in order to do so" (P/K 53). Foucault also confesses, however, to be interested more in using Nietzsche's texts for his own purposes rather than striving for 'correct' and 'faithful' interpretations of Nietzsche's intentions:

> For myself, I prefer to utilize the writers I like. The only valid tribute to thought such as Nietzsche's is precisely to use it, to deform it, to make it groan and protest. And if commentators then say that I am being faithful or unfaithful to Nietzsche, that is of absolutely no interest (P/K 54).

Statements like the last make it a spurious task to try to determine whether Foucault is justified in calling his own enterprise "genealogy" or not. (Or then, it is a task I am not especially interested in.) Thus I shall not attempt here to match Foucault's own writings on Nietzsche against the latter's own texts or against some widely-circulating interpretations. Rather I want to point out briefly, by attending to Foucault's essay "Nietzsche, Genealogy, History", what Foucault finds interesting in Nietzsche's work. Even here, however, caution is called for: given the above quotation, it would certainly be unwarranted to assume that every Foucauldian reformulation of Nietzsche's aphorisms is *eo ipso* a formulation of a thesis of Foucauldian genealogy.

As is obvious enough even from a superficial reading of "Nietzsche, Genealogy, History", there are four main themes in Nietzsche's genealogical writings, that Foucault pays special attention to.

The first theme is that of the contingency of the present, a topic that centers around the notions of struggle and event. Let us call this first theme *the recovering of agonistic events*. Foucault describes Nietzschean genealogy as an attempt to lay bare the contingency of the present and as an opposition to explanations which rely on one exclusive principle of explanation. To do genealogy is to "maintain passing events in their proper dispersion; it is to identify the accidents" of history, it is to understand that "truth or being does not lie at the roots of what we know and what we are, but with the exteriority of accidents" (NGH 81). The aim of genealogy is not to show how the present is necessitated by the past. Rather the aim of the genealogist is to show that the present is the product of a variety of contingent-accidental events of power struggles that could have ended differently from the way they did.[32]

The second theme, an attending to the human body, might be called *physicalism*. Foucault points out that Nietzschean genealogy adopts not only the point of view of the event and the struggle, but also the perspective of the human body as it is formed and molded in these struggles.

The genealogical study of power does not give primacy to the effects of power on the level of beliefs and convictions, but also gives heed to the ways different power regimes control, confine, nourish and shape the human body. Foucault refers approvingly to Nietzsche's idea that genealogy "diagnose[s] the illnesses of the body, its conditions of weakness and strength" (NGH 80). For the genealogist the body is "the inscribed surface of events" (NGH 83), the genealogist does not believe that the human body stands outside of history and that the body is a proper subject matter only for the physiologists. The task Nietzsche sets for genealogy is to "expose a body totally imprinted by history and the process of history's destruction of the body" (NGH 87).[33]

Third, Foucault also shows sympathy for the attitude of *distrust and irony* that informs all of Nietzsche's writings. Allegedly humanitarian motives are to be unmasked as stemming from (and here Foucault quotes Nietzsche) "detestable, narrowminded conclusions. *Pudenda origo*" (NGH 77). The genealogist does not believe that humanity "gradually progress[es] from combat to combat until it arrives at universal reciprocity", rather humanity moves on "from domination to domination" (NGH 85).

This distrust also extends to science and knowledge. Nietzschean genealogy tears off the "the mask" which presents scientific or, more specifically, historical consciousness as "neutral, devoid of passions, and committed solely to truth" (NGH 95). Genealogy seeks to establish that behind this mask all one finds are "the passion of scholars, their reciprocal hatred, their fanatical and unending discussions, and their spirit of competition" (NGH 78), that behind that mask there are but "aspects of the will to knowledge: instinct, passion, the inquisitor's devotion, cruel subtlety, and malice" (NGH 95).

In this unmasking, genealogy takes up an ironic or parodic position. It shows that 'high' motives can be redescribed by allegedly 'low' ones. Thus Foucault can claim that genealogy makes a "parodic and farcial use" of history or that genealogy is "history in the form of a concerted carnival" (NGH 93–94). Genealogy ironicizes the traditional accounts and scientific styles by disassociating itself from them and by showing how they came about. In this sense, the function of genealogy is that of a "curative science" (NGH 90). For instance, in breaking with the traditional view of unity and strong personal identity over time, it opts for the "systematic dissociation of identity": "The study of history makes one 'happy, unlike the metaphysicians, to possess in oneself not one immortal soul but many mortal ones'" (NGH 94).

Fourth, and finally, Foucault also gives much attention to the Nietzschean theme of *perspectivism*. Given the idea that everything is relative to its historical conditions, and given the dissociation of identity, this hardly comes as a surprise. Thus one of the most central traits of genealogy is "its affirmation of knowledge as perspective" (NGH 90): "Truth is undoubtedly the sort of error that cannot be refuted because it was hardened into an unalterable form in the long baking process of history" (NGH 79).

Of course, here Foucault could also have cited Nietzsche's notoriously famous sentences that "facts are precisely what they are not, only interpretations", or that truth is "not something there, that might be found or discovered – but something that must be created ..."[34]

In sum, "Nietzsche, Genealogy, History" suggests that the *recovery of agonistic events, physicalism, distrust and irony* as well as *perspectivism* are the four themes that Foucault regards as of special interest in Nietzsche's genealogy. In the following I shall take up each of these topics in more detail, discussing them from both a historical and a topical point of view.

Recovering Agonistic Events

The return of the event

In order to understand in greater detail what "recovering agonistic events" amounts to as a genealogical research strategy, we have to return once more to Lévi-Strauss and the French historical school around the *Annales*. As seen in greater detail earlier, it was due to the combined efforts of structuralism and *Annales* historians that the notion of event had been a discredited one by the sixties. To begin with structuralism, Lévi-Strauss draws a sharp dividing line between the particular and the universal, and he seeks to leave behind the level of singular historical events qua particulars in order to reach for invariant mental structures qua universals. The latter are modelled upon, or depicted by, means or techniques borrowed from structural linguistics and especially phonology. While Lévi-Strauss leaves a role for history as "indispensible for cataloguing the elements of any structure whatever", history, for him, can be little more than an assistant to anthropology: "... history leads to everything, but on condition that it be left behind" (1966: 262). Anthropology takes over from history in its effort to "grasp, beyond the conscious and always shifting images which men hold, the complete range of unconscious possibilities" (1967: 23). These unconscious possibilities provide us, Lévi-Strauss suggests, with "a logical framework for historical development" (*ibid.*, 24).

In part, the success of the *Annales* school in postwar France was due to its ability to present itself as an ally of structuralist concerns. Lévi-Strauss himself granted that, for instance, the work of Lucien Febvre on Rabelais was congenial to his own work (*ibid.*, 25). Furthermore, a leading exponent of the *Annales* school wrote in 1973, that "for half a century, from Marc Bloch to Pierre Goubert, the best of French historians, systematically systematizing away, have in fact been structuralists. Sometimes they realized it themselves; sometimes they did not; all too often nobody else realized it at all" (Le Roy Ladurie 1981: 5). Moreover, a considerable number of *Annales* historians from Febvre and Bloch on, conceived of anthropology as one of their most important sister discipline. And even those who did not, could at least join forces with structuralism in rejecting the category of event.

By the early 1970s however, several French historians began to raise critical questions concerning the rejection of the category of the event. Questions of priority aside, the work most responsible for a re-evaluation of the notion of event in France, is probably Paul Veyne's *Writing History*, a book first published in 1971. Veyne's thesis is that "history is an account of events: all else flows from that" (1984: 4). For Veyne, historical study is essentially the production of narratives; and here it is the narrative's "plot" that determines what is eventful. While Veyne joins the *Annalists* in criticizing traditional history as a "history of treaties and battles", his criticism turns not against the category of event itself, but rather against the traditional tacit assumption of a set of "'accepted' events", i.e. events that are treated as given. From Veyne's point of view this assumption is mistaken since events are not given but constituted by the historian's ordering of the historical material according to different routes ("itineraries") of inquiry[35]: "In short, the eventworthy field does not comprise spots to be visited and that would be called events; an event is not a being, but an intersection of possible itineraries" (1984: 36).

Against *Annales* historians, on the other hand, Veyne points out that attending to territories, prices and the growth of the population does not inevitably imply a stepping beyond the domain of events. Rather it amounts to a re-ordering of facts according to a new plot and thus to the constitution of new (forms of) events:

> Non-eventworthy history [i.e. *Annales* histories] has been a sort of telescope that, by letting us see in the sky millions of stars other than those known to astronomers of old, would make us understand that our dividing the starry sky into constellations was subjective (*ibid.*, 37).

Foucault's agrees with Veyne (against structuralists and *Annalist* historians) that the category of the event is everything else but obsolete; and, like Veyne, he also points out that rather than exorcising the category of event from history, the *Annales* historians have "constantly enlarged" the field of events by defining ever new series of events (LD 57). However, as in his earlier, archaeological, phase, Foucault acknowledges his indebtedness to the work of Febvre, Bloch and their followers. In fact, in his writings of the seventies he repeatedly expresses his sympathy for, and indebtedness to, the *Annales* school's ideas. With the latter he rejects a history of "those who held power - anecdotal histories of kings and generals" (P/K 51), uses "the great eighteenth-century demographic upswing in Western Europe", much studied by *Annalist* historians, as an explanatory principle for the creation of the modern problems of "population", commends the *Annales* school for its interest in the "'ignorable' materials" of popular culture and the body, endorses their ideal of history as a means of distancing oneself from the present, shares their distrust in the straightforward applicability of linguistic-structuralist models to history, and proclaims himself indebted to their study of spaces (P/K 37, 171; DP 25).

To take up the last two points in more detail, even though members of the *Annales* school like Braudel or Le Roy Ladurie repeatedly express their interest in structuralism and anthropology, they still deny that structuralist methods could be straightforwardly applied to historical materials (Braudel 1980: 44; Le Roy Ladurie 1981: 5). Foucault joins them in this caution, but links it - here of course parting company - to the rehabilitation of the event. For him "the history which bears and determines us has the form of a war rather than that of a language: relations of power, not relations of meaning" (P/K 114). And wars can only be described by giving full heed to the notion of event:

> One can agree that structuralism formed the most systematic effort to evacuate the concept of the event, ... In that sense, I don't see who could be more of an anti-structuralist than myself. ... The problem is at once to distinguish among events, to differentiate the networks and levels to which they belong, and to reconstitute the lines along which they are connected and engender one another. ... 'semiology' is a way of avoiding its [i.e. history's] violent, bloody and lethal character by reducing it to the calm Platonic form of language and dialogue (P/K 114–15).

Foucault's partly critical, partly sympathetic attitude towards "the *Annales* paradigm" is even more clear with respect to spaces, i.e. geographical and architectural spaces. In one passage he claims that it was Bloch's and Braudel's work on rural and maritime spaces and Ariès's work on architecture in the eighteenth century, that decisively opened up the possibility of a study of spaces (P/K 148–49). Still, Foucault suggests that one has to go beyond these studies in so far as the latter are restricted by the conception that space "predetermines a history which in turn reworks and sediments itself in it". In opposition to this view, for the genealogist, space has to be studied as an "economic-political form" (P/K 149). That is to say, architectural and natural space have to be studied in their interrelations with power, with events and processes of struggle.

Recovering contingency

In fact, Foucault's interest in mechanisms and effects of power and the interrelations between power and forms of knowledge is the deeper motivation behind his rehabilitation of the event not only against structuralism but also against the *Annales* school. Explaining his research strategy of "recovering agonistic events" (*eventalisation*) to members of that school in a discussion of *Discipline and Punish* (*QM*), Foucault emphasizes above all three points of difference.

The first ingredient of "eventalisation" is the idea that attending to events rather than structures can help to lay bare the contingency of taken-for-granted self-evidences. The aim of genealogical enquiry is to show that existing structures, like the categorizing of the mad as mentally ill, the hospital system or the prison institution, originate from contingent events and that their existence was thus inevitable only after various processes had taken place (*QM* 104). That is to say, even though their existence became inevitable in some sense at a certain point in history, say *t*, their existence is nevertheless still contingent not only in so far as they owe their existence to a coincidence of various, originally independent, factors, but also in so far as their emergence was unnecessary prior to *t*.

Foucault implies that this contingency of history is not properly attended to by *Annales* historians. He charges them with tending towards replacing historical contingencies by "the most unitary, necessary, inevitable and (ultimately) extrahistorical mechanism or structure available":

> An economic mechanism, an anthropological structure or a demographic process which figure as the climactic stage in the investigation – these are the goals of a history that downgrades events (QM 106; cf. LD 61).

To put the central point more sharply than Foucault does himself, we might want to say that his criticism of *Annales* historians is based on a distinction between events that are "strongly" and events that are "weakly" necessary.[36] An event is strongly necessary if its occurrence at a specific point in the historical course of time had been necessary *all along*; an event is weakly necessary if its occurrence merely *became* necessary *after* certain other events had taken place. Using this distinction between strong and weak necessity, it is inviting to say that Foucault charges the *Annales* historians with paying insufficient attention to the turning points of history, those points *in* or *after* which a structure or institution *becomes* necessary.

Recall also that in "Nietzsche, Genealogy, History" Foucault puts a lot of emphasis on genealogy as the recovering of accidents and chance in history, while in other places he speaks rather of the contingency of the present. Putting these ideas together suggests going a step further and distinguishing between two forms of contingency: contingency-proper or *weak contingency*, and chance or *strong contingency*. To draw this distinction, all we need is a way to drive a wedge between two cases: contingency plus weak necessity, on the one hand, and contingency without any necessity, on the other hand:

> "p is weakly contingent at t" means: first, p obtains at t, and there was some moment of time t', before t, such that at t' it was possible that p would not obtain at t, and, second, there was some moment of time t'', identical with, or after t', such that at t'' it became necessary that p would obtain at t.

> "*p* is strongly contingent at *t*" means: first, *p* obtains *t*, and there was some moment of time *t'*, before *t*, such that at *t'* it was possible that *p* would not obtain at *t*, and, second, there was *no* moment of time *t"*, identical with, or after *t'*, such that at *t"* it became necessary that *p* would obtain at *t*.

The upshot of this exercise, I submit, is this: not only does Foucault want to emphasize - unlike structuralism and the *Annales* School - weak necessity and *eo ipso* weak contingency, but he also tends towards depicting the origin of institutions like the prison as being due to *strong contingency* qua *chance*. To appreciate this point we only have to keep in mind the classical Aristotelian conception of chance as the concurring of two independent causal chains (e.g. I walk to the water because of my thirst, while a group of bandits goes by the same water on their way to town), and Foucault's depiction of, for instance, the origin of the prison as being due to concurring different, partially independent developments, like the new interest in discipline, new architectural ideas, economic problems, new methods of education and factory work.

In sum then, the first ingredient of Foucauldian eventalisation is to suggest a description of those moments of chance or concurrences of different series of events out of which subsequently weakly necessary structures arise.

Causal multiplication

The second ingredient of the Foucauldian research strategy of "recovering the agonistic event", is "causal multiplication" (QM 104). Foucault implies that the historical studies of the *Annales* School are based on a far too limited repertoire of explanatory principles or causes. This charge is expressed not only in the above quotation with its implication that the *Annalists* try to reduce historical plurality to one mechanism or structure; it is also implied in Foucault's explanation as to why his own approach will seem inadequate in this quarter of the historical profession:

> Clearly, viewed from the standpoint of this style of analysis, what I am proposing is at once too much and too little. There are too

> many diverse kinds of relations, too many lines of analysis, yet at the same time there is too little necessary unity (QM 106).

Foucault explains that a singular event or process like the emergence of the prison is a "'polyhedron' of intelligibilities" (QM 105), and that it can be placed in an endless number of series or plots, i.e. in an endless number of relations to other events, none of which renders it necessary.

The agonistic dimension

Finally, the third ingredient of Foucauldian recovery of agonistic events brings in the element of *agon*. In line with his reconstruction of Nietzschean genealogy, Foucault speaks of the importance of rediscovering "the connections, encounters, supports, blockages, plays of forces, strategies and so on which at a given moment establish what subsequently counts as being self-evident, universal and necessary" (QM 105). In other words, the events to be selected by a genealogical study of histories are struggles, conflicts, controversies, i.e. events in and through which new structures become established, existing structures either being used as resources or being destroyed.

Foucault is not alone of course in believing that the study of networks of social power and scientific knowledge, and above all the deconstruction of self-evidences, have to take their starting point from controversies. Such *modus operandi* can be defended by reference to other authors, too. For instance, Shapin and Schaffer[37] in their *Leviathan and the Air-Pump* propose that one of the most effective ways for making the self-evidences of our scientific experimental practices explicit is to take up the position of a "stranger" to this practice. The stranger in a culture becomes easily aware of what are self-evident assumptions for its members simply because the stranger "is in a position to *know* that there are alternatives to those beliefs and practices" (1985: 6). Now one effective means of achieving a stranger's attitude towards our scientific culture, Shapin and Schaffer argue, is to study scientific controversies of the past, controversies out of which this culture emerged. On the one hand, these controversies usually give rise to "disagreements over the reality of entities or properties of practices whose existence or value are subsequently taken to be unproblematic or settled", and, on the other hand, the adversaries in such controversies usually try to "deconstruct the taken-for-granted quality of their antagonists' preferred beliefs and practices ... by trying to display the artifactual and conventional status of those beliefs and practices" (*ibid.*, 7).

However, unlike these authors, Foucault does not limit himself to a study of how structures (e.g. *dispositifs*) and self-evidences (e.g. the normal-pathological divide) emerge from a bundle of disparate controversies and struggles. His call for a recovering of agonistic events, i.e. a conceptualization of history in terms of war, amounts to proposing – in Veyne's terms – a somewhat broader, or longer "itinerary" through the "eventworthy field". Events selected or constituted by this "plot" are, in addition to those out of which (foreseen and unforeseen) structures result, also those events that are predetermined by these structures, and even those events in which these structures and their (foreseen and unforeseen) effects function in multiple, different ways as resources and obstacles for individuals, groups and classes in their struggle for, or against, mechanisms and effects of power and knowledge. For instance, the creation of delinquency was, according to Foucault, an unforseen effect of the penal structure that centrally relied on the prison. Yet members of the bourgeoisie employed this side-effect in order to weaken their primary enemy, the workers, by using the delinquents for the "extraction of profit from pleasure through the organization of prostitution", as provocateurs and as a political-military resource. (Napoleon III rose to power with the help of criminals (DP 280).)[38]

Put differently, on the genealogical plot, controversy does not end with the emergence of structures. Since the structures in question are the result of power struggles, they are always partly unstable and constantly contested. Foucault therefore proposes that history be conceptualized as an ongoing "war", in which "for each move by one adversary, there is an answering one by the other" (P/K 57). In fact, even this formulation is somewhat misleading, since rather than assuming a war between two adversaries, Foucault proposes also that "there aren't immediately given subjects of the struggle, one the proletariat, the other the bourgeoisie". In the war in which ultimately "we all fight each other", there are only ever transitory coalitions (P/K 209).

Foucault is well aware of the fact that his conceptualization of history in terms of struggle is not new as such; foremost it figures of course in Marxism. While acknowledging this, Foucault points out, however, that Marxists have more often than not applied the vocabulary of war and struggle somewhat half-heartedly. He writes, for instance, that "'dialectics' is a way of evading the always open and hazardous reality of conflict by reducing it to a Hegelian skeleton" (P/K 114–15), that "the good old 'logic' of contradiction is no longer sufficient, far from it, for the unravelling of actual processes" (P/K 164) and that even though ...

> the greatest of the Marxists (starting with Marx himself) insisted on the 'military' problems (the army as an instrument of the state, armed insurrection, revolutionary war) ... [when dealing with] ... the 'class struggle' as the mainspring of history, they focus mainly on defining class, its boundaries, its membership, but never concretely on the nature of the struggle (PPHC 123).

This shortcoming of Marxism, Foucault suggests, is to be overcome by replacing the vague notions of "struggle" and "historical contradiction" by a whole network of military terms, and by applying these terms to a wider variety of cases than just the economic sphere. Thus the proponent of agonistic eventalisation does not speak merely, or predominantly, of class-struggle and he does not confine the application of the concept of struggle to instances of military action among states or parties in a civil war. Rather society is a battle-site for power-struggles among individuals, a battle-site for their temporary alliances like interest groups or classes. Each of these individuals or alliances have their own unconscious or conscious strategies and tactics, they try to uphold or bring about a (dis)equilibrium of forces, carry out offensives or counter-offensives along different lines of penetration, retreat or resist or re-organize forces, employ or build *dispositifs* (as ensembles of precautions and measures for carrying out a strategic-military operation), engage in *investment* (as the encircling of a fortress), and strive for their own victory and the defeat of their antagonists.

In sum, then, Foucauldian recovery of the agonistic events suggests – although only in broad outlines – a new plot for the study of social history, the history of science and the history of institutions. To speak of a plot here, and thus make use once more of a category that figures centrally in Veyne's aforementioned study, seems useful because it highlights the fact that this plot functions as a selecting principle within the myriad of historical facts. Needless to say, there are many other historical facts that will inevitably remain outside of the genealogical scope. As a result, agonistic "eventalisation" in particular, and genealogy in general, can make no claims to absolute truth or comprehensiveness. Unlike some other methodologists, Foucault is well aware of this perspectival nature of his enterprise. Thus even though he proposes the employment of this "itinerary" for the study of the history of society, Foucault agrees that it is only one among several ways of cutting up this history. For instance, he allows explicitly for "different strategies (...) for integrating these unbalanced, heterogeneous, unstable, and tense force relations" (HS I 93), and proposes that one is naturally driven towards this plot only when taking the Marxian conception of "struggle" as one's starting point (P/K 164).

PHYSICALISM

The second crucial research strategy arising from Foucault's Nietzscheanism is a focus upon the relations between power and the body. To speak of Foucault's study of the history of the human body as "physicalism" can be justified by more than just his own use of a physicalist idiom when writing about effects of power on the body. More importantly, a specific term for a Foucauldian history of the body seems to be called for, in order to delimit his interest in, and methodology for, this topic from the many other, already existing, "histories of the body".

Histories of the body

Indeed, the historical study of the human body is not as such a new or original endeavour. As Foucault notes himself: "Historians have long ago began to write the history of the body. They have studied the body in the field of historical demography or pathology; they have considered it as the seat of needs and appetites, as the focus of physiological processes and metabolisms, as a target for the attacks of germs and viruses; they have shown to what extent historical processes were involved in what might seem to be the purely biological bases of existence ..." (DP 25).

One can even question whether this list covers the enormous variety of studies that have been carried out, either directly under the heading of "history of the body", or then with obvious relevance to the topic this heading refers to. To give a more adequate overview, the following figure (Figure 14) might be helpful.

Histories of the body investigate how ...

human bodies, their shapes, parts, organs, wants, desires or movements

were ...

sick or healthy, nourished, shaped, distributed, affected by diseases or events in the natural world, treated by oneself or others (trained, punished, observed, studied, distributed, fed, grown, stimulated, dressed, sacrilized, desacrilized ...)

and/or categorized and/or evaluated and/or constituted as ...

normal or abnormal, useful or useless, efficient or inefficient, docile or resisting, male or female, worthy or unworthy, sinful or nonsinful, sexually attractive or unattractive, adult or infantile, physical or spiritual, young or old, strong or weak, alive or dead, sick or healthy, according to race or class, ...

and/or related to ...

other bodies, their shapes, parts, organs, wants, desires or movements, God, the soul, the mind, intentions, ideology, tools, space, time, the state, society ...

and/or how the body or its parts and movements were used as symbols or metaphors for, or interpreted by means of ...

social structures, the natural world ...

by/in ...

individuals, groups, classes, societies, professions, churches, sciences, writers, artists, forms of art, literary genres, periods, social events (e.g. festivals), political events, institutions ...

Figure 14

To flesh out this definition at least briefly, let us note that different traditions and studies have dealt with different aspects or ingredients of this facet sentence. For instance, Robert Mandrou's classic of *Annales* scholarship, *Introduction to modern France 1500–1640. An essay in historical psychology* (Mandrou 1976), opens with two chapters, "The body: food and environment" and "The body: health, diseases and 'population'". These chapters deal with the diet of different social classes, the malnutrition of large parts of the rural population and the diseases resulting from it, clothing, living accommodation, different forms of widespread diseases and the ineffectivity of therapies, over-population, death and birth rates, life expectancy and attitudes towards violence and death. In later chapters, Mandrou also pays attention to emotions, sexuality, conceptions of beauty, manual skills and the representation of the human body in pictorial art. Enquiry into these matters, that also feature repeatedly in the works of other authors of the *Annales* School, are seen by Mandrou as contributing to a reconstruction of the *mentality* of the period under investigation. In other studies, the *Annales* historians have also attended to topics like attitudes towards the bodies of the dead, physical education and corporeal punishment of children (Ariès 1960), or the belief in the Middle Ages according to which touching the King's body can heal one's diseases (e.g. Bloch 1924).

The classical contributions to the history of the body by Norbert Elias and Mikhail Bakhtin, must also be mentioned here. In his *The Civilization Process* (Elias 1978), Elias describes the emergence of the modern "closed" human being. This "*homo clausus*" emerges with the arising of the feeling of embarrassment or the "Peinlichkeitsgefühl" that we experience when coming in close contact with other human bodies, when seeing others spit, eat with their hands, or being nude. Elias argues that the repression of immediate impulses, and the strict demarcation of one's own body from others, was part and parcel of the individualism and the respect for the other sex that slowly emerged in the Renaissance. Elias also shows that this rise of "civilization" was linked to the coming of the modern national state.

Bakhtin's book, *Rabelais and His World* (Bakhtin 1968), can be read as depicting the kind of body that fell prey to the historical processes described by Elias. Bakhtin investigates what he refers to as the "carnevalesque" or "grotesque" body of the lower classes in medieval times and the Renaissance, a body that is spontaneous, laughing, disrespectful, and intimate with others.[39]

Finally, and to bring this incomplete list to a provisional end, one must also mention the numerous studies on the history of medicine and on images of the body as involved in social and political thought. For instance, studies carried out in and around the Edinburgh School in the sociology of knowledge have frequently attended to the ways theorizing in medicine, phrenology and eugenics related to political events and class-struggles. For example, the 18th century assumption of feelings of sympathy, an assumption made by physicians to explain the integration of the body, was used by the Edinburgh elite as an analogy of society, and as a resource for justifying the political *status quo* (Lawrance 1980).

Body, power, knowledge

While there certainly are links between these various studies of the human body, on the one hand, and the Foucauldian "physicalism", on the other hand, the latter also has several characteristic features that are worth attending to in more detail.

Foucault himself sees the novelty of his project in its being based on the observation that ...

> the body is also directly involved in a political field; power relations have an immediate hold on it; they invest it, mark it, train it, torture it, force it to carry out tasks, to perform ceremonies, to emit signs (DP 25).

Foucault's project in the study of the body is thus an investigation into the relations between the body and power, or, more precisely, the relations between social power, (scientific) knowledge, the human body and personal identity.

This historical-political study of the body is motivated, above all, by two ideas. On the one hand, it seems natural - given Foucault's activities on behalf of prisoners, the mentally ill and hospitalized people - to attribute to him the view that it is first and foremost the concrete physical suffering of human beings that a critical reflection of modern society must attend to both politically and by way of historical study. On the other hand, Foucault's interest in the triad of body-knowledge-power is also motivated by a criticism of the prioritizing of "ideology" in Marxist analyses of society. In one place, Foucault suggests that it might be "more materialist" to study the effects of power on the body rather than to focus exclusively on questions of ideology (P/K 58). In other places, he goes further and questions the usefulness of the notion of ideology altogether. Thus he claims, first, that the concept of ideology presupposes a problematical opposition with truth, second, that it is based on a traditional notion of the subject, and third, that it is taken as distinct from, and seen as determined by, an economic basis or infrastructure (P/K 118). Rather than assuming naively à la Althusser the existence of a non-ideological, scientific truth as the opposite of ideology, Foucault holds that we have to study how "effects of truth are produced within discourses"; rather than presupposing "a human subject on the lines of the model provided by classical philosophy, endowed with a consciousness which power is then thought to seize on" (P/K 58), we have to study the creation or formation of this subject; and rather than trying to explain all power as ultimately deriving from the economic infrastructure, we have to diversify our explanatory models (P/K 118).

The deeper reason why Foucault finds an analysis in terms of ideology questionable, and why he also denies the link between his physicalism and the study of mentalities, is his belief that - especially in the modern world - mechanisms of power have evolved that do not necessarily involve conscious, mental representations in the power subject. Instead, these new forms of manipulation make the body move in specific ways without the power subject being able or forced to deliberate on these movements. Moreover, given these new forms of manipulation, even if the power subject deliberates on her actions, this deliberation has already been influenced and shaped by systems of control and surveillance over her body.

Since the details of this analysis belong more properly to a critical review of Foucault's historical investigations, let us turn to his conceptual apparatus for investigating the relation between power and bodies.

Now, it is obvious that coercion has a rather intimate relation to the body. Not only are many coercive threats directed against the body of the power subject, but the power subject is also usually directly physically confronted with the coercer. Both coercer and coerced are typically *seen* by one another, and of course the coercer must make the coerced believe that any deviating from the demanded course of action will be observed immediately. Furthermore, every modern society has a system of coercive institutions, like the police or the prison, a system to guarantee and enforce the punishment of lawbreakers. And, needless to say, a considerable number of punishments are corporal, in that they confine the body to a cell, threaten it with bodily pain, or destroy the body's life.

Foucault's investigations into the penal systems of different historical periods suggest that these can be studied as so many "regimes" or "modalities" of bodily confinement and the infliction of pain. Generalizing one of the notions that Foucault only applies to the modern form of power over the body, we might say that every society has a specific "political anatomy" (DP 138), a specific way of relating to the bodies of different groups, sexes, classes, and age groups, and that this "anatomy" comes out most clearly in the way society punishes members of these different categories.

Some passages of *Discipline and Punish* also suggest that these regimes can be measured along various axes. A first such axis is the distinction between two directions of visibility: some systems of power rely more on the threatening visibility of the power holder, others employ techniques of surveillance that leave the power holder by and large invisible.[40] Second, systems of control also vary as to their "scale" (DP 136) or degree of "resolution" (P/K 151): some force and bind the body as a whole, as in the case of slavery, while others exercise "upon it a subtle coercion, ... obtaining it at the level of the mechanism itself – movements, gestures, attitudes, rapidity: an infinitesimal power over the active body" (DP 137). Third, power over the body can be exercised by inflicting pain, by training or by stimulating it.[41]

For Foucault, various systems of "political anatomy" need to be studied by attending to the kinds of knowledge or truth they presuppose and produce. In fact, any study of the body-power relation that leaves out this further variable of knowledge is insufficient from the standpoint of genealogy, since it is, as will be remembered, one of the most basic assumptions of the Foucauldian theory of power, that power is always interrelated with knowledge. Neither can it be claimed that the interrelation between power over the body and knowledge is merely a peculiarity of modern societies with their specific forms of surveillance and control. After all, to mention only one example that *Discipline and Punish* discusses in some detail, judicial torture as the inflicting of extreme pain on the suspect's body, obviously was, and was conceived of as, a search for truth (DP 37).

The genealogical questions with respect to the triad of knowledge-power-body fall, roughly, into three groups.

The first is concerned with the know-how or 'practical' knowledge involved in systems of bodily control and punishment. This knowledge is a "technological" knowledge of power mechanisms, a knowledge of measures by which bodies are most effectively treated, by which pain is most efficiently applied, and by which the body is best trained, disciplined, stimulated and controlled. This technological knowledge is knowledge on how to intensify power over the body, how to lower the costs of power exercises, how to standardize the means of power, and how to increase the amount of power. It is knowledge of how the power of the body *qua* capacity is increased and of how the power of the body *qua* resource of resistance is diminished (DP 138). Of course, this knowledge is not only presupposed by systems of bodily surveillance and control, it is also refined and developed within such systems by trial and error.

Foucault usually speaks of this "political technology of the body" (DP 26) when dealing with the specific form of technological knowledge involved in more recent forms of disciplining. However, his own study of the knowledge of bodily pain involved in earlier systems of torture suggests that this term can also be applied more widely. What makes this generalization even more natural, is Foucault's view that this knowledge is below the threshold of systematic science, to wit, that it is "diffuse" ...

> ... rarely formulated in continuous, systematic discourse; it is often made up of bits and pieces; it implements a disparate set of tools or methods. In spite of the coherence of its results, it is generally no more than a multiform instrumentation. Moreover, it cannot be localized in a particular type of institution or state apparatus. For they have recourse to it; they use, select or impose certain of its methods (DP 26).

The second group of genealogical question with respect to the body, power and knowledge, focusses on scientific knowledge about the body, its movements and functions. Genealogy asks how this scientific knowledge is related to systems of control and the technological knowledge behind them. Here Foucault's well-known thesis is that knowledge of the human sciences as we know them, and especially their methods, were made possible in part by "the great nineteenth century effort in discipline and normalization" (P/K 61). This is especially clear in the case of medical knowledge, one condition of which were new methods of observation and control employed in the hospital. Here poor patients were - and still are - used for medical experiments, and patients are constantly put under surveillance. For instance, they are distributed according to their symptoms and diseases in order to facilitate the accumulation of medical truths. Furthermore, often scientific knowledge of the body feeds back into the systems of control and technological knowledge from which they emerge; cases in point are doctors working as consultants in factories and prisons, as torture-instructors, or as technical advisors in executions (PPHC 193).[42]

Adding to Foucault's suggestions, we can also note that again we are not dealing with a totally new phenomenon: the interrelation between power, body and knowledge is not confined to the modern era; after all, cruel medical experiments on the living bodies of convicts were performed way back in the third century B.C. in Alexandria by Herophilos and Erasistratos - not even to mention the myriads of animals that have been tortured to death over the last two millenniums in the name of scientific progress.[43]

Third, and finally, the genealogical study of the history of the body as caught in the networks of science and social power, also questions the body in its status as a prediscursive, natural object. That is to say, Foucault seeks to lay bare the ways in which the body, its pleasures and its sexuality have been constituted by systems of power and knowledge. In other words, Foucauldian physicalism, here once more quite congenial to Faraday's enterprise, does not take bodies as given, and does not allow for any clear line between the biological and the historical:

> ... what is needed is to make it [i.e. the body] visible through an analysis in which the biological and the historical are not consecu-

> tive to one another, as in the evolutionism of the first sociologists, but are bound together in an increasingly complex fashion in accordance with the development of the modern technologies of power that take life as their objective. Hence I do not envisage a 'history of mentalities' that would take into account of bodies only through the manner in which they have been perceived and given meaning and value; but a 'history of bodies' and the manner in which what is most material and most vital in them has been invested (P/K 152).

To tackle this problem is, for instance, to study the emergence of the "sexualized bodies" (and the differences between male and female bodies) in and through the interplay of various discourses (sexology, medicine, eugenics, demography, pedagogy, jurisprudence, psychiatry, obstetrics ...), methods of control and surveillance (in boarding schools, birth clinics, army camps ...), as well as techniques of stimulation (pornographic literature, films ...).

Distrust and Irony

Turning finally to the third main genealogical research strategy, we can start from the observation that skepticism with respect, for instance, to the notions of man, progress or mental illness is of course a well-known feature of Foucault's work throughout its different stages. More generally, Foucault cautions us against the tendency to assume that concepts or words like 'madness' or 'deviance' denote natural objects; he asks us to see these objects as resulting from prior objectivations.[44] I shall be dealing with this skepticism or attitude of suspicion in the next chapter under the heading of Foucault's "(methodological) perspectivism". Here I rather want to take up Foucault's Nietzschean *Leitmotif* of distrusting, parodizing or ironicising self-proclaimed purely scientific and humanitarian interests or motivations, i.e. his attention to non-scientific interests in analyses of the historical emergence of the human sciences and coercive institutions.

For instance, cases in point are Foucault's arguments to show that the criticism of the prison reform movement in the 19th century was less motivated by humanitarian concerns than by the interest to bring about a more effective penal system[45], and Foucault's attempts to demonstrate that the control of children's sexuality was as much motivated by doctors' and parents' interest in control and sex as it was motivated by medical concerns.[46]

Interests

Foucault's emphasis on such non-humanitarian and non-scientific interests and power structures is of course congenial with recent work in the sociology of science, especially the studies carried out by the Edinburgh School. To focus more closely on these parallels, suffice it here to take a closer look at the lecture "The dangerous individual" (DI). In this paper Foucault's strategy of explaining scientific and social development via interests can clearly be seen.[47]

In "The dangerous individual" Foucault investigates why crime in the 19th century became increasingly medicalized or connected to mental disease. He argues that a first step in this direction was taken when, in the 1830's and 40's, psychiatrists began to publish widely on the phenomena of "the ultimate crime", i.e. cruel, terrible murders allegedly without visible motivation or reason. Psychiatrists in part helped to create, in part enrolled, the publics' as well as the judges', awareness for these crimes, offering their own profession of psychiatry as promising to provide the key to their solution.

Psychiatrists exploited the public fear of, and the judges' perplexity over, these crimes by introducing the notion of "homicidal monomania", i.e. the notion of a form of insanity whose only visible expression or symptom is homicide. As Foucault puts it: "Nineteenth-century psychiatry invented an entirely fictitious entity, a crime which is insanity, a crime which is nothing but insanity, an insanity which is nothing but crime" (DI 132). More precisely, the nature of this form of insanity was claimed to be such that only the psychiatrists could eventually come to recognize its almost invisible symptoms.

It was by redefining the ultimate crime, a crime linked inextricably to death and thus to fear, that psychiatrists went a long way in convincing the public as well as many legal experts of their competence in the realm of crime and punishment. It was by redefining the cruel, motiveless homicide in psychiatric terms that psychiatrists gained a position of social power that they had not had before.

However, Foucault cautions us not to attend only to the professional vested interests of the psychiatrists: "It is not enough to invoke some sort of imperialism on the part of psychiatrists seeking a new domain for themselves ..." (DI 133). In order to understand why they could succeed, one has to take into account the earlier success of the medical profession in convincing politically powerful groups, like industry, the army and the police, of its own importance as an expert in public hygiene, i.e. as an expert in treating of the diseases of the newly invented category of the "population". Psychiatrists thus did not have to start their own struggle for a larger share of social power from scratch. The medical profession, from which psychiatrists had just evolved as a more or less independent field, had provided the latter with a strategically important basis. Psychiatrists merely needed to extend the new interpretation of the doctors' socially crucial role to a new domain.

Moreover, the psychiatrists' interest in penetrating the legal apparatus was not independent of their other professional interests. The attention of psychiatrists to this whole issue was indeed connected to their concern for justifying their right to impose a therapeutic confinement upon the mentally ill. Homicidal mania merely worked as one case where the public's fear and the judges' helplessness could be translated into the need for such confinement.

Why then were the judges unable to defend their traditional power of being the sole adjudicators of punishment? This too is an issue Foucault proposes to address in terms of power. He argues that during the 18th century the traditional system of punishment as based on revenge and deterrence had become inefficient as a system of power and had been replaced by a new system based on panoptical surveillance and reforming of the offender. The new focus on the individual, a focus that came hand in hand with the new power mechanisms of surveillance, confession and discipline, shifted awareness from the criminal act to the criminal individual. In determining punishments, the judiciary was thus forced to determine a punishment by considering how dangerous the individual in question was and how he or she could be re-educated. However, precisely in the case of the unmotivated, unreasonable murderer this was impossible. In short, the psychiatrists had chosen the right target for their attack.

The notion of the dangerous individual which was thus emerging could then be exploited by the anthropology of criminal man in Italy and the theory of social defense in Belgium. Once psychiatrists had succeeded in firmly establishing their hold on the legal apparatus towards the end of the 19th century, they no longer needed the notion of homicidal monomania. In fact, they could - and had to - give up this notion, in order to further expand their area of influence. Relying on a more refined picture of mental disease as developing by degrees, on the one hand, and on the notion of degeneration, on the other hand, psychiatrists could establish their professional competence with respect to ever more forms of criminality. Psychiatry "enlarged, organized, and codified the suspicion and the locating of dangerous individuals, from the rare and monstrous figure of the monomaniac to the common everyday figure of the degenerate, of the pervert, of the constitutionally unbalanced, of the immature, etc." (DI 149).

Even this brief sketch of a much more involved argument suffices for us to see that Foucault's treatment of interests and power has strong similarities with the use of interests as explanatory devices in the Edinburgh type of sociology of science. Interests focussed upon in the latter are, foremost, (1) "professional nested interests" as scientists' interests to (re-)define object areas for the purpose of establishing their (exclusive) competence in dealing with them; (2) scientists' interests in keeping laymen out of the scientific field; (3) scientists' interests in using conceptual tools drawn from the broader cultural or social context (in order to arouse wider attention for their work); (4) interests in the wider society to make scientific results effectively applicable; and (5) scientists' interests in drawing support from classes or institutions (Shapin 1982).

All of these interests are also touched on by Foucault in various of his works, even though not always under the same heading of interests. It is especially interesting to note that Foucault agrees with these sociologists not only in claiming that straightforward explanations in terms of classes are often rather questionable, but also in proposing that professional nested interests play a crucial role and that scientific work often is dependent upon the support it manages to draw from state institutions, industry or the Church.

More important than these parallels, however, is the fact that

Foucault's account summarized above is not open to one specific charge that has been raised against the Edinburgh program. According to this criticism, authors of the Edinburgh School neglect the fact that the interests of the wider society which scientists enroll for their purposes are often not pregiven, but rather produced by the scientists themselves. This point has been especially convincingly treated by Bruno Latour in his study of Pasteur (Latour 1983; cf. Latour 1987, 1988). For him, the art of using interests is the art of "convincing others of what their interests are and what they ought to want and to be. He who is able to translate others' interests into his own language carries the day" (1983: 144). In Pasteur's case the interest of the public was "a consequence and not a cause of Pasteur's efforts to translate what they want or what he makes them want. They have no a priori reason to be interested at all, but Pasteur has found them more than one reason" (*ibid.*, 144).

Of course, a similar line is involved not only in Foucault's dealing with the arising of the category of "the dangerous individual", but also in *Discipline or Punish* or the *History of Sexuality I.* After all, the way to success for psychiatrists was to translate the interest of the public to be freed from the fear of death and crime into the interest both in the confinement of various allegedly insane individuals and in the work of psychiatrists themselves. The psychiatrists had of course helped to bring about this fear by writing extensively on these special types of crime.

There is also a noteworthy parallel between Latour's rejection of the distinction between the inside and the outside of the laboratory and Foucault's attention in the two-way borrowing of mechanisms of power (and the knowledge coming with them) between coercive institutions and the society at large. Latour deconstructs the border between the laboratory and the wider society by arguing, first, that experimentators typically translate "outside" problems into laboratory settings, and, second, that the application of laboratory results to the outside world is, in fact, an extension of laboratory conditions to the outside: "... on the condition that you respect a limited set of laboratory practices – disinfection, cleanliness, conservation, inoculation gesture, timing and recording – you can extend to every French farm a laboratory product made at Pasteur's lab" (*ibid.*, 152). In Foucault, the Latourian laboratory is replaced by coercive institutions, but otherwise the line of thinking is similar: coercive institutions are the social laboratories for new forms of control and discipline (and the new forms of knowledge that support, or arise from, these power mechanisms) and the results of these social experiments are extended to ever new social domains. For instance, Foucault argues that the effectiveness and docility of the worker is achieved by transposing the power mechanisms of the prison and the army to the workplace.

We might perhaps also claim that Foucault holds a counterpart of Latour's thesis that "most of the time when we talk about the outside world we *are simply taking for granted the prior extension of a former science* built on the same principle as the one we are studying" (*ibid.*, 156). Translated into Foucault's framework this would amount to the claim that the outside world is but a network of power, in part modelled on some earlier coercive institutions. In any case, these two French authors certainly concur that "in our modern societies most of the really fresh power comes from sciences - no matter which - and not from the classical political process" (*ibid.*, 168).[48]

Ironic hypotheses

Replacing the standard picture of, say, psychiatry as motivated through and through by a humanitarian concern, with a picture of psychiatrists' as being motivated by interests in power and manipulation, has, of course, something of the burlesque, the parodic or the ironic. After all, what Foucault - and the sociologists - are, in the eyes of many scientists and philosophers, engaged in, is an undue replacement of the dignified by the 'low' and 'familiar'. Hardly surprising then that their writings and ideas should meet with so much resistance.

Foucault has been aware of this ironic or parodic element of his work from the beginning; not unexpectedly for a student of Nietzsche rather than of Mannheim, Merton or Habermas.[49] This element is rather obvious not only in some of Foucault's asides or interviews ...

> Why should so many people, including psychiatrists, believe that I am an anti-psychiatrist? It's because they are not able to accept the real history of their institutions which is, of course, a sign of psychiatry being a pseudo-science. A real science is able to accept even the shameful, dirty stories of its beginning (PPHC 15).

... but also in Foucault's research strategies. After all, Foucault asks us to be aware of developments in and through which attempts to increase the efficiency and amount of power backfire: the medical discourse on homosexuality provides homosexuals with a vocabulary to articulate their rights; the controlling of sexuality leads to its stimulation; and the controllers become themselves controlled (the Panopticon). These invisible hand effects, by themselves, provide a burlesque commentary on the work of the architects of power.

In any case, it seems fair to say that Foucault favors the methodological strategy of attempting to proceed by using, what one might term, 'ironic hypotheses'. When confronted with an institution whose openly proclaimed purpose is to bring about, or hinder, or uphold a set of states of affairs, say *S*, Foucault will suggest working on the opposite, ironic, assumption that the real purpose of the institution was the hindering of *S* (rather than bringing *S* about), or the bringing about of *S* (rather than its hindering), or changing *S* (rather than upholding it). Just think here of Foucault's hypotheses that the prison's function was to produce rather than prevent crime, or that the medical classification of perversions was an incentive rather than a rejection of these 'deviations'.

Finally, we might note also the reflexivity of Foucault's irony, i.e. his conscious attempt to make the reader aware of the transient and provisional character of his own models and conceptions. Foucault's writings do not employ a single framework, but from *Madness and Civilization* up to the latter parts of the *History of Sexuality*, we are in fact confronted with several vocabularies. Moreover, as seen earlier, Foucault uses various different models for conceptualizing power, and some of them are clearly incompatible (war, politics, physics, individualist). Thus, for Foucault, there is not *one single* framework for the study of power, but rather there are "different strategies (...) for integrating these unbalanced, heterogeneous, unstable and tense force relations" (HS I 93). Perhaps Foucault's reminder that "I have never written anything but fictions", a reminder that is qualified by the addition that "I do not mean to say, however, that truth is therefore absent" (P/K 193), is best understood in this sense, too.

13. GENEALOGICAL PERSPECTIVISM

INTRODUCTION

The fourth Nietzschean *Leitmotif* of Foucault's genealogical research, i.e. perspectivism, is important and complex enough to be best treated in a separate chapter. Admittedly, perspectivism is a rather vague notion. Even in the case of Nietzsche himself, interpreters use this concept in different ways. While they agree that for Nietzsche all knowledge is relative to perspectives, interpreters disagree over the implications of this view. According to some Nietzsche scholars, his perspectivism does not rule out the possibility of speaking meaningfully of truth and of evaluating different perspectives in some non-arbitrary fashion (Nehamas 1985: 64–73). Others hold that Nietzsche rules out this possibility: for them, perspectivism amounts to the doctrine that since everything is relative to perspectives, we should not even use the concepts of truth and falsehood (MacIntyre 1988: 352).

In this chapter, I shall use perspectivism as an umbrella-notion. I outline Foucault's position with respect to such issues as constructivism, skepticism and relativism, and then simply label his specific views on these notions "perspectivism".

In dealing with relativism, skepticism, and constructivism, we enter a rather complex and emotionally loaded area of philosophical debate. What makes a discussion of these concepts difficult is not only the vagueness of their meaning, but also, and following from this vagueness, the fact that the positions these concepts denote seem praiseworthy to some philosophers, but utterly distasteful, self-defeating and dangerous to others. Often, for philosophers who take the latter view, there is not much, for instance, that distinguishes relativism and skepticism from, say, "total skepticism", "nihilism", or "irrationalism".

In suggesting that Foucault adopts some versions of relativism, constructivism and skepticism, I thus *prima facie* discredit his work in the eyes of more than just a few readers. However, rather than joining them in rejecting Foucault's thought, I shall try to convince them that relativism, constructivism and skepticism are not absurd positions. In particular, I shall try to argue that epistemological relativism is not a self-refuting doctrine, that methodological anti-realist constructivism is not self-defeating, and that skepticism is coherent in both a limited and an unlimited form.

SKEPTICISM

Foucault has been alternately praised and attacked for being a skeptic. For instance, John Rajchman calls Foucault "the great skeptic of our times" and endorses Foucault's thought as one that aims "for the freedom of withholding judgment on philosophical dogmas, and so of acquiring relief from the restrictions they introduce into our lives and our thought" (Rajchman 1985: 2). Some critics of Foucault, however, see his skeptical position as self-refuting in so far as they regard his alleged denial of all knowledge as pre-emptying the possibility of claiming truth for his own work (e.g. Merquior 1985: 147; Merquior 1986: 259).

Against the latter position it is worth stressing that *even if* Foucault were to be the universal skeptic these critics make him out to be, his position *need* not be self-refuting. Indeed, it is possible to provide a description of genealogical studies that avoids the charge of self-refutation with respect to those studies. Foucault could be a universal skeptic and still write historical studies meant to undermine, case by case, the central standards and categories of philosophy, psychiatry and the human and social sciences. If we take his studies as being addressed to the 'dogmatic' proponent of these standards and categories, their employment by Foucault would not involve him in a contradiction: his use of these standards and categories would only be a hypothetical and tentative one. Foucault's arguments, premisses and conclusions, would be arguments, premisses and conclusions *for the proponent* of these standards, meant to convince these proponents that it is best to withhold judgement on these standards.

However, charges of self-refutation aside, Foucault's skepticism is not the universal skepticism that some of his critics accuse him of. To see this, we only have to note, for example, that Foucault never, neither directly nor implicitly, suggests that there is no knowledge, no progress, and no truth. To be sure, Foucault does endorse skepticism in the following passage:

> I adopt the methodological precaution and the radical but unaggressive skepticism which makes it a principle not to regard the point in time where we are now standing as the outcome of a teleological progression which would be one's business to reconstruct historically: that skepticism regarding ourselves and what we are, our here and now, which prevents one from assuming that what we have is better than - or more than - in the past (P/K 49).

Note, however, that Foucault hastens to add: "I don't say that humanity doesn't progress. I say that it is a bad method to pose the problem as: 'How it is that we have progressed?'" (P/K 50).

Furthermore, Foucault's genealogical texts abound with defenses of "subjugated knowledge" as well as with claims like "I know that knowledge can transform us", "knowledge is for me that which must function as a protection of individual existence", and "I'm an empiricist" (PPHC 7, 14, 106).

Foucault's skepticism is thus at most a limited one. His skepticism is first and foremost a skepticism with respect to the notions that have been used for writing the history of the last three-hundred years. These are the notions of an ever-increasing welfare and freedom, of the growing humanitarian care for the sick and the mad, of the education of ever wider strata of the population, of our freeing ourselves from the suppression of sexuality, of the emergence of a scientific spirit that increasingly frees itself from structures of power and domination, and of a liberating philosophical discourse.

Foucault's limited skepticism in these mentioned cases does not amount to denying that progress has, in fact, taken place. Rather his skepticism grows from the disquieting awareness that, to put it in terms of Wittgenstein's favorite quotation from Nestroy, "it is the nature of all progress that it looks much greater than it really is"[50]. Foucault - like e.g. Adorno, Horkheimer and Wittgenstein - stresses that each progress has demanded a high price, and that often the progress has been measured one-sidedly. For instance, while it is true that the application of the death penalty has been abandoned in some Western countries, to praise this as a victory of humanitarianism is to overlook that the death penalty is still applied on a large scale in the United States as well as in more than one-hundred countries around the world. Furthermore, in the

West, the extensive application of capital punishment in the case of a few major crimes has been replaced by a new system of punishment which has enormously increased the range of punishable offenses and has introduced large-scale imprisonment. As documented by Foucault and others, these developments have led, and still lead, to the permanent exclusion of large numbers of individuals from society.

To repeat, if Foucault can be said to suggest a withholding of assent, this withholding does not amount to a withholding of assent with respect to all knowledge claims. It would be more accurate to say that Foucault wishes to withhold assent or unqualified approval with respect to the standard narrative of progress, and, especially, with respect to the normative conceptions that come with this plot.

Foucault's skepticism can be seen most clearly in his treatment, or non-treatment, of normative ethical or political notions, a fact that reflects itself in Foucault's approval of the label of a "skeptical ethics" (1977: 67) for his work. Since he claims to have shown that many of the notions that are used to justify progressive movements have actually been shaped by the very "regimes of truth" and "regimes of practices" they are meant to undermine, he wishes to abstain from using or endorsing them in a systematic manner. Rather than building on these traditional standards of ethics and politics, Foucault follows the classical skeptic in turning to the immediate experience: in his case, to the suffering of the allegedly mad, the prisoners, and the victims of medical tutelage.

CONSTRUCTIVISM

The second key ingredient of genealogical perspectivism, "constructivism", brings genealogy into close contact with more recent relativist sociology of science. Two forms of constructivism, i.e. the doctrine that scientific facts result from human activity, can be distinguished. "Realist constructivism" holds that even though scientific facts are "remade" in, say, the laboratory, these facts are still "discovered" and determined by reality. According to "irrealist constructivism", scientific facts are *only or merely* the result of human activity. Facts are nothing but fabricated, and they are not discovered. Nature is the result of construction, and puts no constraints on beliefs and theories (Hacking 1988; cf. Knorr-

Cetina 1983). For the irrealist, it makes no sense to speak of nature or reality as determining an experimental outcome, since the very categories of nature or reality are themselves taken by the irrealist to be mere products of scientific negotiations. As Bruno Latour and Steven Woolgar, two somewhat ambivalent proponents of this position, put it:

> "Reality" cannot be used to explain why a statement becomes a fact, since it is only after it has become a fact that the effect of reality is obtained (1986: 180).
> It is not simply that phenomena *depend on certain material instrumentation; rather the phenomena are thoroughly constituted by* the material setting of the laboratory. The artificial reality, which participants describe in terms of an objective entity, has in fact been constructed ... (*ibid.*, 64).

In other places, however, Latour and Woolgar bend more towards realistic constructivism; e.g. at one point they write that "we do not want to say that facts do not exist and that there is no such thing as reality" (*ibid.*, 180).

The leading figure of the Bath School in the sociology of scientific knowledge, Harry Collins, urges sociologists of science to adopt the methodological standpoint that "the natural world in no way constrains what is believed to be" (1982: 140). Here nature has dropped out completely as a determining factor of knowledge, and as the determining factor of experimental outcomes.

Collins calls his own approach "relativism". This is adequate, since he, like Latour and Woolgar, and other sociologists of science more or less loosely associated with the Edinburgh School, sees scientific knowledge as relative to the partly differing standards of various scientific cultures or groups. Be it noted, however, that irrealism, as formulated by Collins above, need not go hand in hand with epistemological relativism. Even if reality does not determine our beliefs, it is still possible that there are universal standards of rationality or ways of categorizing which determine a singular "true" conception of the world. Analogously to the distinction between universalist and relativist nominalism, one might thus also distinguish between a universalist and a relativistic irrealism. This move is inviting too because nominalism can be looked upon as a special case of irrealism: irrealism with respect to categorizations.

To show that Foucault adopts irrealism (and thus *eo ipso* nominalism) at least as methodological points of view, is not especially difficult. After all, already in his archaeology Foucault studied how objects like madness are constituted in and through a number of contingent processes. In other words, madness is not something waiting 'out there' in nature or society to be discovered, but it can be discovered only once it has been constituted in, by and through various practices and struggles.

Foucault also states straightforwardly in several passages that his interest is not to study how nature determines our knowledge, but rather to investigate the processes of how certain 'natural' objects or categories come into existence in and through both discursive and non-discursive practices. Recall, for instance, his earlier quoted insistence that we "should try to discover how it is that subjects [and by the same token, the categorizations of subjects] are ... constituted through a multiplicity of organisms, forces, energies, materials, desires, thoughts etc." (P/K 97). He also tells us that we must study scientific discourse as "a practice that we impose upon" nature (LD 55), and that truth is to be studied not so much as a correspondence with reality but rather as "a system of ordered procedures for the production, regulation, distribution, circulation and operation of statements" (P/K 133). Thus where the sociologists of science wish to study how statements put forward in the laboratory are eventually turned into facts such that "all traces of production are made extremely difficult to detect" (Latour and Woolgar 1986: 176), Foucault is involved in a similar project with respect to central concepts and "facts" of human "nature". Furthermore, where Latour and Woolgar seek to demonstrate "how a hard fact can be sociologically deconstructed" (*ibid.*, 107), Foucault attempts a similar feat by genealogical-historical means.

While there thus can be little doubt about the question whether or not irrealism is part and parcel of genealogy, it is worth pointing out that Foucault is careful not to propose irrealism as a global standpoint with respect to all sciences. He seems in fact to take irrealism foremost as a methodological, heuristic device, i.e. he studies practices and scientific discourses *as if* in their constituting activity these were not constrained by nature. Foucault neither claims that scientists and social practices are free to construct any object or category they like, nor does he ever imply that the construction of madness is not – in some way or other – related to brain-damage and types of unusual behavior (cf. Veyne 1979: 47). The reason why he need not take these 'objective facts' into account from his starting point should be clear enough: in the discursive and non-discursive interactions that Foucault studies, these 'objective' features inevitably appear only as already interpreted.

To be sure, if Foucault were to endorse irrealism on a total scale, then his own studies could be nothing but fairy-tales: if what is 'out there' in nature and society does not determine our theories and beliefs at all, we would have no reason to believe Foucault's own story of what is 'out there' in society and history.

Fortunately, Foucault's position is, to repeat, more moderate. What his methodological irrealism amounts to is merely the Canguilhemian idea according to which "the object of the history of science [as a discipline] has nothing to do with the object of science". That is to say, the object of the history of science is science as a social community or as a succession of bundles of theories, whereas the object of science is nature or society. While genealogy and the sociology of science bracket the latter, they do not bracket the former. In other words, the genealogist does not leave "the natural attitude" when studying the processes in and by which science and apparatuses constitute objects. But he gives up this natural stance and practices an *epoché* (in both the skeptical and Husserl's sense) with respect to those objects that the scientist takes for real.

It is also possible, of course, to iterate this move from the natural attitude to the genealogical one. We can not only study how sciences constitute *their* objects; we could also investigate how, in turn, genealogy constitutes *its* subject matters of disciplines and apparatuses. The only move that is indeed a likely candidate for impossibility is the attempt to do both investigations at one and the same time. Be it said also, to conclude, that the infinite regress which opens up here can hardly be used as an argument against the genealogical enterprise: after all, the distinguishing feature of human action and thinking is precisely the fact that it inevitably contains or implies reflexivity.[51]

Epistemological Relativism

Before taking up the question of what the third ingredient of genealogical perspectivism, Foucault's epistemological relativism, amounts to, I want to address the more basic question of whether charging a thinker with being an epistemological relativist is to provide a decisive argument against him or her. The reason why the vast majority of philosophers and social scientists would answer in the positive is, obviously, the widespread belief that epistemological relativism is a self-refuting, incoherent standpoint. In this and the following section, I wish to suggest that this received view is mistaken.

In order to establish this position, I shall, admittedly brief, take a closer look at two recent books, one by Harvey Siegel and one by Alaisdair MacIntyre, books that set out to refute epistemological relativism. Of course there are numerous other authors who have attempted a similar feat. Restricting myself to these two writers can be justified, however, by noting that since Plato there have been few innovations in the reasoning against relativism and that the arguments presented by Siegel and MacIntyre can be seen as culmination points of this long tradition of 'refutations'.

"Relativism Refuted" – refuted

Siegel, in his book *Relativism Refuted. A Critique of Contemporary Epistemological Relativism* (1987), takes epistemic relativism (= *ER*) to be the following doctrine:

ER: For any knowledge-claim *p*, *p* can be evaluated (assessed, established, etc.) only according to (with reference to) one or another set of background principles and standards of evaluation $s_1, \ldots s_n$; and, given a different set (or sets) of background principles and standards $s'_1, \ldots s'_n$, there is no neutral (that is, neutral with respect to the two (or more) alternative sets of principles and standards) way of choosing between the two (or more) alternative sets in evaluating *p* with respect to truth or rational justification. *p*'s truth and rational justifiability are relative to the standards used in evaluating *p* (*ibid.*, 6).

Let us look more closely at Siegel's three main arguments against *ER*. The first two arguments reformulate, in more exact terms, ideas already employed by Plato's Socrates against Protagorean relativism.

(1) Both Plato and Siegel hold that epistemic relativism is incoherent "because if it is right, the very notion of rightness is undermined, in which case relativism cannot be right" (*ibid.*, 4). Siegel's point is that the non-relativist is entitled to demand from the epistemic relativist "right reasons" for the latter's position, where by "right reasons" Siegel understands reasons that are not relative, i.e. reasons based on a "neutral (with respect to the presuppositions of relativists and non-relativists) framework" (*ibid.*, 8). Siegel's twist then is simplicity itself: if the relativist provides these "right reasons", she has *ipso facto* transcended his relativism and thus has refuted herself, since, according to epistemological relativism, there are no non-relative reasons. If, on the other hand, the relativist does not provide any non-relative, but merely relative reasons, the absolutist is not obliged to adopt relativism, since for her, "'relative rightness' is no rightness at all" (*ibid.*, 8). Siegel holds that relativism, in taking all standards to be relative, makes it impossible to provide any objective non-relative standards in terms of which epistemological positions - and among them relativism - can be evaluated.

It is fairly easy to show that Siegel's first argument fails to demonstrate the incoherence of epistemological relativism. No relativist needs to be impressed by Siegel's line of reasoning, for all it shows is that relativism allows only for relative reasons, and that relativism is the denial of the existence of a universally valid framework. But to show this is not to refute relativism, it is merely to restate its content. To answer the relativist challenge, it is insufficient to tell the relativists what they already have been saying all along; if the absolutist wants to refute this position she better point out to the relativists just what a list (preferably a longish list) of non-relative grounds for deciding epistemological issues looks like. This, however, Siegel fails to do.

Let us note also that Siegel's claim, according to which relative rightness is no rightness at all, is not very convincing. For instance, right moves in the game of chess are certainly right only relative to the game, yet they are right all the same.

Finally, we can also point out against Siegel that the giving of relatively-right arguments for relativism is not as odd and insufficient as he claims. Arguments for relativism are not *per se* "impotent", and adopting the relativist position need not be the totally arbitrary decision that Siegel makes it out to be. ("... the relativist can adopt beliefs (and act) only arbitrarily" (*ibid.*, 21).) Given that both the framework or culture relativist and her critic, i.e. the absolutist, share the same framework and culture, arguing for the relativist position can have force, and adopting the relativist stand is not merely up to the individual. Given the shared framework (shared with their absolutist opponent) the relativist can, without contradicting herself, present arguments. Of course, the consensus possibly emerging on this basis of a common framework will be a local one, and perhaps only temporary. But this is not in itself an argument against relativism, especially since no transcendental arguments for necessary *a priori* truths and standards have so far gained general acceptance.

(2) Siegel's second argument turns on the idea that "necessary some beliefs are false": "relativism is incoherent because it holds that all beliefs and opinions are true, yet, given conflicting beliefs, some beliefs must necessarily be false – in which case relativism cannot be true" (*ibid.*, 6).

The details – and mistakes – of this line of reasoning can be spelled out best by attending closely to Siegel's reformulation of Plato's way of putting this argument against Protagorean relativism. The central passages from Siegel's text are the following (I have labelled the sentences for subsequent reference):

> [a] Protagorean relativism is false for all those who do not believe it: [b] it is false for all if no one believes it; and [c] true only to the extent that some number of people (perhaps only Protagoras himself) believe it ... [and Siegel quotes Plato:] [d] "it is more false than true by just so much as the unbelievers outnumber the believers" ... [e] if opinions conflict about the truth of Protagoras' relativism, then the Protagorean relativist must acknowledge the truth of the opinion that the doctrine is false" (*ibid.*, 5).

Applying this argument to his reformulation of *ER*, Siegel writes that ...

[f] it follows that, if according to some set of standards $s_1, \ldots s_n$ ER is judged to be false, then, if ER is true, (at least according to that set of standards $s_1, \ldots s_n$) ER is false. In this way, ER is self-refuting, and so incoherent (*ibid.*, 6).

To see why this line of reasoning is mistaken, be it noted first of all with respect to Siegel's formulation of the second argument, that relativism need not hold that all beliefs and opinions are true. Within a given framework, the distinction between opinion, belief and knowledge, or the distinction between truth and falsehood can be maintained. Siegel's way of formulating the relativist stand is thus inadequate, to say the least.

Second, note that in moving from [a] to [d], Siegel drops the relative notion of truth as truth-for-someone, and replaces it with the absolute notion of truth. This is a question-begging move since the relativist will precisely deny that truth as a non-relative, absolute notion, makes sense. Siegel (and Plato) cannot, in order to avoid circularity, neglect this position, since their aim is to derive an incoherence *from premisses accepted by the relativist.*

To make the *petitio principii* of Siegel's reasoning explicit, let us reformulate his and Plato's statements in more exact terms.

Siegel's [a] ("Protagorean relativism is false for all those who do not believe it") is the claim that

[a] $(x)(\neg B_x[PR] \rightarrow \neg T_x[PR])$
(For all persons x: if x does not believe in Protagorean relativism, then Protagorean relativism is not true-for-x.)

This statement can be accepted by the Protagorean relativist. However, Siegel's [b] ("it is false for all if no one believes it") has two readings. The reading that restates [a] would be [b']:

[b'] $(\neg Ex)(B_x[PR]) \rightarrow (x)(\neg T_x[PR])$
(If there is no person x, such that x believes in Protagorean relativism, then for all x: Protagorean relativism is not true-for-x.)

On the second reading, a reading that must be at work in order to justify Siegel's subsequent claim [c], truth is not relative:

[b"] $(\neg Ex)(B_x[PR]) \rightarrow (x)(B_x \neg T[PR])$
(... then for all x: x believes that *PR* is not true.)

Only given the equivocation that exists between [b'] and [b"] does it become understandable that Siegel can move on to the claim that [c] (*PR* is "true only to the extent that some number of people [perhaps only Protagoras himself] believe it") ...

[c'] $(Ex)(B_x[PR]) \leftrightarrow T[PR]$
(*PR* is true if and only if there are some persons x such that these x believe in *PR*.)

[c'], with its unindexed use of the truth operator, obviously flies in the face of the basic contention of relativism, whether Protagorean or not. Even worse, let us note that Siegel's choice of words invites yet another equivocation. His expression "true only to the extent" might also be read as an allusion to degrees of truth:

[c"] $(Nx)(B_x[PR]) \leftrightarrow d(T;N)[PR]$
(Protagorean relativism is true to a degree, if and only if there is a number of people believing it; where any specific degree of truth is dependent upon specific numbers of believers.)

That it is not a wild speculation to attribute this reading to Siegel is obvious from the fact that he approvingly quotes Plato's [d] ("it is more false than true by just so much as the unbelievers outnumber the believers"). Formally:

[d] $(d(T)[\neg PR] > d(T)[PR]) \leftrightarrow ((Nx)(\neg B_x[PR]) > (Nx)(B_x[PR]))$
(The degree of truth of not-*PR* is greater than the degree of truth of *PR*, if and only if the number of x not believing *PR* is greater than the number of people believing it.)

It hardly needs to be explicitly stated at all that nothing of this sort is claimed by either Protagoras nor the modern relativist. The dropping of the indices is also to be lamented upon in [e] ...

> Given that *a* believes in *PR* ($B_a[PR]$) and that *b* does not ($\neg B_b[PR]$), all that *a* is forced to acknowledge is that *PR* is not-true-for-*b* ($\neg T_b[PR]$). *a* is not forced to admit, however, that *PR* is not true ($\neg T[PR]$), and even less is *a* obliged to regard *PR* as not-true-for-herself ($\neg T_a[PR]$). From '*a* believes that *b* does not believe in *PR*' ($B_a \neg B_b[PR]$), or from '*a* believes that *b* believes *PR* to be false' ($B_a B_b[\neg PR]$), it does not follow that *a* must abstain from believing *PR* ($\neg B_a[PR]$).

... and [f]:

> From 'epistemological relativism is true' ($T[ER]$) and epistemological relativism is not true for a framework *fr1*' ($\neg T_{fr1}[ER]$) it does not follow that 'epistemological relativism is not true' ($\neg T[ER]$), simply because the relativist does not accept the unindexed notion of truth, i.e. she does not accept $T[ER]$ as a well-formed formula. All that follows is, trivially, that 'epistemological relativism is not true for framework *fr1* ($\neg T_{fr1}[ER]$). Again, given that two frameworks differ in their truth value assignments with respect to *ER* ($\neg T_{fr1}[ER]$ and $T_{fr2}[ER]$), it does not follow that *ER* is false ($\neg T[ER]$) or that it is false for both frameworks. At most, what might follow is that the relativism-endorsing framework has to allow for the possibility of *ER* being false when judged on the basis of other frameworks. But that the assessment of epistemological theses varies with frameworks, is part and parcel of the epistemological relativism anyway. How could *ER* itself be exempt from it?

(3) Siegel's third argument seeks to establish that the very notion of relative truth does not make sense and that relative truth presupposes an absolute notion of truth. Siegel stresses this point when attacking Jack Meiland who has proposed that truth-for-*W* (where *W* can range over frameworks, cultures etc.) "denotes a special ... relation which does not include the ... relation of absolute truth as a distinct part" (*ibid.*, 13). Now, Meiland puts this position somewhat clumsily by saying that "one

can no more reasonably ask what 'true' means in the expression 'true-for-*W*' than one can ask what 'cat' means in the word 'cattle'" (*ibid.*). Countering this example, Siegel has no difficulty in convincing his readers that ...

> 'cat' is no more a meaningful part of 'cattle' than is 'ca', 'catt', cattl', etc. 'True-for-W', on the other hand, is made up of the hyphenization of distinct concepts ... (*ibid.*, 13–14).

However, Siegel's suggestion for a better example, to wit, "action-at-a-distance", will not do either since it prejudges the issue in his favor. For given this example, it is easy to claim that truth-for is but a special conception of truth (*simpliciter*). After all, action-at-a-distance falls under the concept of action (*simpliciter*).

To make Meiland's position immune to Siegel's attack, it suffices to say that what the proponent of "truth-for-*W*" is in fact doing is not so much defining a new concept of truth as pointing out a hidden variable which has been at work in the "absolute" concept of truth. Introducing the hyphenization of "truth-for-*W*" is like pointing out to a child which uses the words "left" and "right" in an absolute sense, that the real meaning of "left" is "left-of", and the real meaning of "right", "right-of". Obviously, correcting the child in this way does not presuppose any "absolute" notion of leftness. In other words, the relativist holds that in the absolute notion of truth one *relatum* has been overlooked, or that one variable has been held constant, to wit, our ways of modelling, conceptualizing and conceiving of the world. When couched in these terms, it is easy enough to appreciate that the concept of relative truth is not completely different from the traditional notion of truth, and that Siegel's criticism of Meiland's "truth-for-*W*" (viz. "he has provided no hint what the string does mean or refer to" (*ibid.*, 14) does not carry the day. Telling a child that "left" really means "left-of" does not, of course, mean that one has introduced something completely new. One is merely pointing out to the child a feature of the concept that the child had formerly overlooked. *Mutatis mutandis*, the relationship between "truth" and "truth-for-*W*" is not such that the latter is parasitic upon the former.

The "not-to-be-neglected truth"

MacIntyre's attitude towards relativism seems to be, at least *prima facie*, more sympathetic than that of Siegel. For MacIntyre, relativism contains a "not-to-be-neglected truth". An indication for the existence of this truth, MacIntyre suggests, is the fact that relativism has been refuted so often in the history of philosophy, whereas "genuinely refutable doctrines only need to be refuted once" (1987: 385). This "not-to-be-neglected truth" is assumed to reside in the feeling of incomprehension that we experience when encountering the texts of other cultures (*ibid.*, 404).

In his recent book, *Whose Justice? Which Rationality?* (1988), MacIntyre also gives more general definitions of relativism and perspectivism. Relativism is the doctrine that ...

> to assert or to conclude this rather than that can be rational relative to the standards of some particular tradition, but not rational as such. There can be no rationality as such. Every set of standards, has as much and as little claim to our allegiance as any other (*ibid.*, 352).

However, perspectivism goes further than relativism by drawing a radical conclusion from the relativist stand: we should not use the notions of truth and falsehood (at least in their traditional sense), and we are to treat of different traditions or frameworks as so many perspectives "for envisaging the realities about which they speak to us" (*ibid.*, 352).

MacIntyre presents us with two arguments against relativism (and thus *ipso facto* against perspectivism) and an additional criticism against perspectivism in particular.

According to the first argument, relativism neglects the possibility that within every tradition the debate over, or the development of, a "problematic" can reach the point where members of the tradition begin to question the standards of that tradition, and where ultimately the claim of that tradition to truth can no longer be upheld:

> And this by itself is enough to show that if part of the relativist's thesis is that each tradition, since it provides its own standards of

rational justification, must always be vindicated in the light of those standards, then on this at least the relativist is mistaken (*ibid.*, 364).

MacIntyre's second argument continues along the same line. It cannot be true that rationality equals the rationality of any particular tradition, he argues, because within a tradition one can come to see that the only way forward in dealing with its "problematic" is to turn to other, rival, traditions and their standards (1987: 408; 1988: 364). This undermines the relativist position, MacIntyre alleges, because relativism claims that "there is no way in which each tradition can enter into rational debate with any other, and no such tradition can therefore vindicate its rational superiority over its rivals" (1988: 366).

Finally, and this is the argument specifically addressed to the perspectivist, MacIntyre proposes that the perspectivist neglects the importance and inevitability of the notion of truth for traditions of enquiry. In other words, we cannot give up this notion because it is "integral ... to tradition-constituted forms of enquiry" (*ibid.*, 367).

As is easy enough to show, there is something suspicious not only about MacIntyre's three arguments, but also about his definition of relativism. As we already pointed out when criticizing Siegel's arguments, for the relativist, the choice of standards and the adoption of beliefs need not be totally arbitrary. Unfortunately, this is, nevertheless, what MacIntyre assumes ("Every set of standards has as much and as little claim to our allegiance as any other.")

MacIntyre's first argument will work only if the relativist denies the possibility of a development of standards within traditions or frameworks, and if the relativist disallows for the possibility that one can question some of a given tradition's standards in terms of other standards of that same tradition. However, all the relativist needs to deny – in order to be a relativist – is that there is some neutral ground of universal standards, a ground in terms of which the standards of a given tradition or framework, or then the failure of some of these standards, can be assessed and evaluated.

Concerning MacIntyre's argument from encounters between traditions, three points must be stressed. First, MacIntyre himself concedes that even though a tradition might 'objectively' need the help of another tradition, "... it does not of course follow that there will be actual acknowledgement" (*ibid.*, 365). Seeking help from the other thus seems to be a demand of "rationality ... *qua* rationality" (1987: 408). To see that this will not do, one only needs to echo the title of MacIntyre's book: Whose rationality? If this rationality is taken to be a universal one, then the question has been begged. If it is the rationality of some tradition, however, the relativist position has not been undermined.

Second, if MacIntyre wants to claim that the possibility of one tradition's learning from another does presuppose some common criteria or standards, then this, in itself, is not sufficient as a refutation of the relativist. Even if common standards were needed in this case, all the relativist needs to reject is that these criteria or standards can be universal, or *a priori* necessary. But this specific denial does not rule out the possibility of there being standards which are relative, and common, to at least two traditions.

Third, it also is not clear why the relativist would have to refuse to acknowledge that the standards, the knowledge and the truths of some given tradition, say tradition *A*, can be judged as superior with respect to another tradition, say *B*. All the relativist has to contradict is that this evaluation can be made from a ground which transcends all traditions. But this leaves open the possibility of this evaluation being made from within *A* or *B*, or both, or from within some other tradition.

Finally, MacIntyre's point against perspectivism (truth as inevitably presupposed within traditions) does not succeed either, since – according to his own definition of this stand – the perspectivist only rejects truth and falsehood in their traditional sense, not in every sense.

To conclude this brief discussion of whether or not relativism is a self-refuting doctrine, it remains to be said that even *if* the above arguments against epistemological relativism as a theoretical position *were* to succeed, this would still not endanger it as a heuristic device or methodological stand. In other words, even the absolutist might accept the idea that working on the basis of the relativistic hypothesis, i.e. that knowledge is nothing over and above what is accepted as such within a given culture (be it scientific or not), is a more fruitful hypothesis for historical and sociological research than working on the foregone conclusion concerning the existence of universal principles of rationality, common to all (possible) cultures. Perhaps, the absolutist might even adopt this position as a methodological rule in the hope of collecting material which, when day dawns, will undermine epistemological relativism as a theoretical stand.

Foucault's relativism

In showing that epistemological relativism is not the self-defeating, impotent and arbitrary view it is often made out to be, I hope to have given the reader reason to believe that the epistemological relativism of Foucauldian genealogy provides no compelling reason for rejecting the Foucauldian project. What remains to be done here is merely to present the details of genealogical relativism.

A first feature of genealogical epistemological relativism is that the truth-defining frameworks and relative standards are essentially related to social power. Foucault speaks of "truth" as a set of rules and prescriptions for the production of scientific propositions, and he refers to these sets as "regimes of truth" in so far as these knowledge-defining frameworks are related to the systems of social power and sustain, reinforce, and profit from, the knowledge produced (P/K 133).

Here it is worth emphasizing once more, and contradicting critics of Foucault[52], that the essential-internal link between truth and social power does not as such turn truth into arbitrary opinion and illusion. On the contrary, it is Foucault's very point that every socially organized production of truth and knowledge is possible only against the background of, and within, a framework of rules, practices and standards, to wit, that the production of knowledge is possible only against the background of, and within, systems of social power. (As seen earlier, however, Foucault considers that not all of these systems of social organization are evaluatively on one par.)

Second, Foucault does not hold[53], that regimes of truth are homogeneous, that they cannot be changed from within, and that they cannot be compared. That regimes of truth are not homogeneous is clear enough from Foucault's constant reminders that within the same large-scale system of social power, i.e. within the same "strategy", several scientific and philosophical discourses can exist, side by side, and that these discourses abide, at least in part, by different, even incompatible standards:

> Discourses are tactical elements or blocks operating in the field of force relations; there can exist different and even contradictory discourses within the same strategy; they can, on the contrary, circulate without changing their form from one strategy to another, opposing, strategy (HS I 102).

Foucault also writes that "relations of power-knowledge are not static forms of distribution, they are 'matrices of transformations'" (HS I 99). Furthermore, he suggests that even though different systems of domination have their own specific rationalities, "transmissions, transferences, interferences" also occur (PPHC 37). Moreover, just as power in general leads to effects which undermine its hold and encourage opposition to it, so do scientific discourses. Foucault proposes, e.g. that even though the new discourses in the nineteenth century on "the species and subspecies of homosexuality, inversion, pederasty, and 'psychic hermaphrodism'" made possible and led to an increase in social control, the very same discourses also led to the formation of an opposing discourse: "homosexuality began to speak in its own behalf, to demand that its legitimacy or 'naturality' be acknowledged, often in the same vocabulary, using the same categories by which it was medically disqualified" (HS I 101). Finally, Foucault also seeks to rescue "subjugated knowledge" as knowledge or beliefs that do not qualify as true by the standards of a given "regime of truth". In sum, it is hard to see how Taylor can maintain that "Foucault tidies it up too much, makes it [i.e. history] into a series of hermetically sealed, monolithic truth-regimes" (Taylor 1986: 98).

In connection with the problem of how existing regimes of truth can be changed or questioned, Foucault maintains that we can "turn the will to truth ... against truth" (LD 22). In other words, even though there is no vantage point from which we could bystep our framework or regime of truth, we still can investigate (on the basis of *our* form of knowledge) into our framework's origin and historically contingent nature. The aim of such inquiry, we are told is to "free thought from what it silently thinks, and so enable it to think differently" (HS II 9):

> What reason perceives as *its* necessity, or rather, what different forms of rationality offer as their necessary being, can perfectly well

> be shown to have a history; and the network of contingencies from which it emerges can be traced. Which is not to say, however, that these forms of rationality were irrational. It means that they reside on a base of human practice and human history; and that since these things have been made, they can be unmade, as long as we know how it was that they were made (PPHC 37).

Naturally, this does not mean that we can come to know our own framework in full from within itself, nor that we can find ourselves outside of every framework. On the one hand, the possibility of coming to know our own framework completely is denied where Foucault writes that "we have to give up hope of ever acceding to a point of view that could give us access to any complete and definite knowledge of what may constitute our historical limits" (ENL 47). On the other hand, Foucault explicitly rejects the possibility of any universal standards; he speaks of rationalities in the plural, and suggests that there is no vantage point from which our forms of rationality could be regarded as absolutely superior to earlier ones (PPHC 36). He also proposes, for instance, that the ceremony of public torture was not irrational in its time, even though (most would say) it is irrational for us today. In other words, there is no "absolute against which [different systems of punishment] could be evaluated as constituting more or less perfect forms of rationality" (QM 107).

Clearly, Foucault's suggestion that we study the origins of our standards and frameworks historically as emerging from earlier ones further undermines the claim that he conceives of different frameworks as being completely incomprehensible to one another. For example, even though he holds that the Greeks "had their own "regime of truth" (PPHC 223), he still sees nothing strange about a study of Ancient Greek morals, a study that is hoped to lead one "to explore what might be changed, in [one's] ... own thought, through the practice of a knowledge that is foreign to it" (HS II 9).

To conclude, while being relativistic in the sense of denying any universal standards of rationality, and while holding that rational judgments can only be made on the basis of historically contingent standards, Foucault does not adopt any of the implausible ideas that MacIntyre and Siegel see as definitive of relativism and perspectivism: for Foucault, the historical nature of our standards of knowledge does not make our standards – for us – impotent; he gives no "absolute" reasons for relativism, confining himself instead to a case-by-case demonstration of how allegedly timeless categories have had historically contingent origins; he does not make the truth or validity of standard a function of the number of frameworks abiding by them; he does not base his relativism on any claims of incommensurability and intranslatability; he does not reject the use of the concepts 'true' and 'false', and he does not exclude the possibility of incoherent standards within the same framework.

14. GENEALOGICAL CRITICISM OF POWER AND RATIONALITIES

Perhaps the most striking difference between Foucauldian genealogy and research programs in the sociology of science is that genealogy does not confine itself to descriptions and explanations of power and interest structures in science. Rather, genealogy takes up a critical posture towards the power of science in modern society. To analyse this posture in a separate chapter seems imperative for two reasons. On the one hand, it allows us to see genealogical methodology at work, and illustrates the critical accumen of genealogical relativism. On the other hand, the overwhelming majority of critics of genealogy have concentrated their attention on the critical posture of genealogy. To answer their criticism will, I hope, turn the focus of debate over genealogy away from this topic, and towards the issues dealt with in previous chapters.

Genealogy as Criticism and Analysis of Rationalities

Let us start by reviewing Foucault's characterizations of the role of genealogical analyses concerning specific political issues.

Foucault suggests that since the late sixties a series of oppositions and struggles has emerged: women have begun to question male dominance more radically, children question the authority of their parents, and psychiatry is criticized for its control over the mentally ill, medicine for its treatment of the sick, and administrations and bureaucracies for their interventions in the realm of the individual (S&P 211). Foucault proposes that these resistances have a number of common traits, such as their international character, their being conflicts over power effects, their immediacy (rather than being struggles against a class or the state, they are struggles against specific effects), their being concerned with the right of individuals to determine their own identity, and their being struggles against privileges of knowledge (S&P 212). These resistances are all similar in that they involve the rehabilitation of forms of subjugated knowledge and in that they are intertwined with theoretical-intellectual work (P/K 81).

Obviously, the issues listed remind one of the topics that Foucault deals with in his genealogical writings, especially if one remembers that the problem of sexuality is linked to the male-female and the parents-children question, and that the effects of discipline and stimulation (dealt with in *Discipline and Punish*) are related to the working of administrations and bureaucracies.

But just what contribution can Foucauldian genealogy make to these local struggles and resistances? To put it succinctly, Foucault believes that the specific critical contribution of genealogy is not so much, or not primarily, to criticize institutions and persons, but instead to *make criticizable* the forms of knowledge, the standards of rationality, or principles of reasoning not only on the basis of which such institutions arise and turn out to be 'reasonable', 'useful' and 'self-evident', but also on the basis of which the violence of these institutions becomes natural and justifiable. Thus genealogy provides the groups opposing a given institution with an additional critical weapon: the debate over the institution need no longer be carried out in terms of those forms of knowledge or frameworks of reasoning which underlay the institution itself.[54]

As Foucault points out time and again, his aim is first and foremost the "wearing away [of] certain selfevidentnesses and commonplaces", the changing and displacing of "ways of perceiving and doing things ... or ... of forms of sensibility and thresholds of tolerance" (QM 109), the grasping "why and how that-which-is might no longer be that-which-is", the criticizing of "political rationality's very roots" and the making intelligible in order to make criticizable (PPHC 36, 85, 101). That is to say, rather than analyzing the working of factually existing institutions, genealogy seeks to identify the network of knowledge, the standards of rationality or rationalities that make these institutions possible and that justify them. Since it is the central premiss of genealogy that knowledge and power are internally-essentially related, genealogy cannot remain on the level of violence and coercion in institutions like the prison; it has to study the frameworks of thinking and knowledge that inform these institutions.

The stress upon the analysis of rationality *in the plural* is crucial here. Foucault tells us that to work with the simple opposition between reason and unreason, or with the notion of a bifurcation of reason into emancipatory and technical-strategical, is unhelpful. On the one hand, relying on the dichotomy 'reason versus unreason' forces us to say either that the penal system and its justification is based on reason or that it is based on unreason. In the first case, we as critics place ourselves outside of rationality and thus on the side of irrationality. In the second case, we fail to realize the reasoning underlying the institution, and thus underestimate its specific rationality and effectiveness. In brief, "such a trial would trap us into playing the arbitrary and boring part of either the rationalist or the irrationalist" (PPHC 59). On the other hand, relying on the dichotomy of technical-strategic versus emancipatory reason, an idea central for the Frankfurt school, tends to make us blind to the variety ("an endless, multiple bifurcation") of different forms of strategic thinking, and reinstalls a simple-minded opposition between reason and power. (PPHC 29, 35).

What genealogy attends to then is not whether people who rationalize something "conform to principles of rationality, but to discover which kind of rationality they are using" (PPHC 59). Perhaps we might say that Foucault's epistemic relativism shows its critical acumen here: the idea is to show on the basis of which – historically contingent – principles of reasoning certain strategies of control and punishment are rationally defensible. Of course, to show this is to endorse neither these strategies nor their principles. Instead it is to open up a new level of debate, a level that lies deeper than the immediate criticism. What genealogy strives to do is to demonstrate that *even though* certain judgments and measures, say within the penal system, are defensible given the underlying historically contingent principles and (local) frameworks, these same judgments and measures are *only* defensible given these historically contingent principles and (local) frameworks. To restate a point stressed earlier, this is not to say that these frameworks or principles can be questioned from some point beyond all frameworks, but it is to show that these frameworks are not without alternatives, that we can modify or enlarge them, that we can question them as merely local frameworks from within the more global framework of our culture.

The task is thus, as Foucault puts it, "a liberation of thought", that is meant to prepare the ground for "deep transformations" rather than small corrections within a given system (PPHC 155). To take the issue of penal institutions, this liberation is one that allows us to question not only specific insufficiencies in prisons, but rather to question the current system of penal justice as a whole. By the same token, we acquire the means "to avoid other institutions, with the same objectives and the same effects, from taking their stead" (PPHC 84).

Foucault characterizes this radical form of criticism in a discussion in 1971, a discussion in which he opposed his own form of work to what he calls "the humanist's" critique. He claims that whereas a humanist critique takes as given that, "the guilty are guilty and the innocent are innocent", and merely confines itself to demanding minor improvements in the prison for the prisoner as a human being ("... consequently, flush toilets"), genealogy "isn't concerned with the soul or the man *behind* the convict, but it seeks to obliterate the deep division that lies between innocence and guilt" (LCMP 227).

After the publication of *Surveiller et punir*, Foucault writes in a somewhat less radical tone. In 1977 he says that the aim of genealogy's analysis of those rationalities in which the current penal system is embedded is to make small, straightforward adjustments within the prevailing system difficult:

> ... it's true that certain people, such as those who work in the institutional setting of the prison – which is not quite the same as being in prison – are not likely to find advice or instructions in my books that tell them 'what is to be done'. But my project is precisely to bring it about that they 'no longer know what to do', so that the acts, gestures, discourses which up until then had seemed to go without saying become problematic, difficult, dangerous (QM 110).

In other words, genealogy does not only abstain from making proposals for limited reform, but even brings into question these very standards by which we would be likely to evaluate proposals as to their efficiency, humanity, or inevitability. For Foucault, to justify one's criticism with the putting forward of a program for reform is to fall into "a trap in which those who govern try to catch intellectuals" (PPHC 52); to fall into this trap is to accept the validity of the idea "don't criticize, since you're not capable of carrying out a reform" (QM 110). The reason why this challenge can be rejected is not only that any such immediate and straightforward proposal for reform must remain bound to the prevailing standards, but also because those who govern usually withhold from their critics the very information necessary for deciding which solutions are possible (PPHC 52), and also, most importantly, because "'what ought to be done' ought not to be determined from above by reformers, be they prophetic or legislative, but by a long work of comings and goings, of exchanges, reflections, trials, different analyses" (QM 110). Put differently, political decisions on the penal system, for example, should be based on experimentation and a prolonged debate between not only philosophers and the legislators, but with all the parties, including the prisoners, who are involved. In this experimentation and in this debate genealogical analysis participates only by laying bare the contingency of the standards of those who see the present system as sufficient and defensible.

Is Resistance Possible?

The doubts that critics of Foucault have raised concerning the critical posture of genealogy fall roughly into two groups. On the one hand, many writers have claimed that, given Foucault's theory of power, all criticism of modern science and society is futile. This is so, these writers allege, because the latter theory leaves no room for resistance to power structures. On the other hand, critics hold that the absence from Foucault's writings of a normative theory for organizing a just society leaves him without a normative basis for any criticism. In the reminder of this chapter, I shall take up these two issues in turn.

To begin with the possibility of resistance, the thesis that I wish to attribute to Foucault is the following: *While power relations are an inevitable fact of social life, not all power relations are evaluatively on a par: we can distinguish – roughly – between hegemonial forms of power which we can resist and non-hegemonial forms of power which we need not resist.*

While Foucault is notorious for his reluctance to make value judgments, it is interesting that in several contexts he in fact draws a distinction between inevitable or harmless, and avoidable or harmful forms of power. Thus, even though he says that we cannot ever reach a state where truth is emancipated "from every system of power (which would be a chimera, for truth is already power)", he calls for the "detaching [of] the power of truth from the forms of hegemony, social, economic and cultural, within which it operates at the present time" (P/K 133). Foucault also tells us that genealogy fights for ...

> the insurrection of forms of knowledge that are opposed primarily not to the contents, methods or concepts of [...] science, but to the institution and functioning of an organized scientific discourse within a society such as ours (P/K 84).

The underlying premiss of these statements is formulated by Foucault himself as: "Every power relation is not bad in itself, but it is a fact that always involves danger" (PPHC 168).

The fact that Foucault believes it is possible to resist hegemonial power is obvious not only from the task that he assigns to genealogy, but it is also clear from many other contexts. For instance, Foucault holds that we can speak of a power relation only where there is the possibility of "refusal or revolt", even though the latter might consist only of preferring death over submission. Foucault thus draws a distinction between power and violence. Violence or force leave the victim no choice between compliance or non-compliance; in these cases the victim is treated as a body and not as someone whose deliberation is to be influenced. ("... [power] is a mode of action which does not act directly and immediately on others. Instead it acts upon their actions. ... A relationship of violence acts upon a body or upon things" (S&P 220).)

In one passage, Foucault calls the element of resistance inherent in power relations "a certain plebeian quality or aspect" (P/K 138). While doubting that the plebs is a genuine sociological entity, and while denying that the plebs can be identified with the proletariat, Foucault suggests that the plebeian quality is "a centrifugal movement, an inverse energy, a discharge" and that it can be found all over the social body "in a diversity of forms and extensions, of energies and irreducibilities":

> This measure of plebs is not so much what stands outside relations of power as it is their limit, their underside, their counter-stroke, that which responds to every advance of power by a movement of disengagement (P/K 138).

While I certainly agree with Foucault's critics that Foucault's conception of the resisting agent is underdeveloped, and that his histories of the prison and the asylum give little attention to resistance, it still seems to me that the often pronounced worry, according to which Foucault's theory of power makes resistance inconceivable, is vastly exaggerated.[55]

First, the claim that power is omnipresent clearly does not lead to this result. From the fact that our daily lives are permeated both by different, mostly harmless, forms of exercises of power, and by various forms of power being exercised over us, it hardly follows that we cannot oppose, e.g., the extensive power that professors exercise over unemployed young scholars, or the extensive power that the Finnish police has over foreigners living in Finland. Critics tend to misread the Foucauldian idea of the inevitability of *some or other* forms of power as the much stronger claim that *all* of the *now*-prevailing power relations are inevitable and thus irresistible. But attributing the latter position to Foucault is to fall prey to the fallacy of composition.[56] It is to infer *B* from *A*:

(*A*) For all times and all parts of society: there exist *some (or other)* relations or mechanisms of power.

(*B*) For *some (and the same)* relations or mechanisms of power, these relations and mechanisms of power exist for all times and all parts of society.

The latter statement is contradicted not only by Foucault's stress on the historical development of power mechanisms, but also by his allowing for resistance against malicious forms of power.

Actually, one might even argue that the omnipresence of power makes resistance more rather than less likely. After all, this thesis includes the idea that all of us are exercisers as well as victims of power, coercers as well as victims of coercion. And this can only mean that none of us has fully adopted a submissive attitude that predisposes us to be a willing victim of coercion: all of us harbour some of the rigidity which makes us less likely to give in.

Furthermore, we also need to give proper heed both to Foucault's notion of subjugated knowledge, and the contradictory effects, or unforseen by-products, of mechanisms of power. Recall that Foucauldian subjugated knowledge includes the knowledge of those – the prisoner, the mentally ill, the patient – who more than any group in the Panoptical society are the victims of power mechanisms. The fact that hegemonial forms of power can only suppress but not destroy these forms of knowledge shows that even the most massive manipulation cannot, on Foucault's view, determine the victim's identity completely. In part, this situation is due to the circumstance that no coercive institution, and no system of coercive institutions, is ever immune to the danger of producing or using several, conflicting mechanisms of power such that they either neutralize one another or have other unforeseen and resistance-invoking counter-effects. Foucault explicitly allows for the possibility that the combining of infinitesimal mechanisms leads to "numerous phenomena of inertia, displacement and resistance" (P/K 142). He also indicates, more concretely, that the preoccupation with the health of the human body, a preoccupation central to the theme of bio-power, has in fact led to such countereffects:

> ... once power produces this effect there inevitably emerge the responding claims and affirmations, those of one's body against power, of health against the economic system, of pleasure against the moral norms of sexuality, marriage, decency. Suddenly, what has made power strong becomes used to attack it (P/K 56).

Finally, the claim that Foucault does not conceive of human subjects as being through and through determined by manipulation and control, can also be backed by attending to a notion that surfaces in his last writings, *The History of Sexuality II* and *III*. This notion is that of "technologies of the self", a notion a discussion of which lies beyond the scope and central interests of my investigations. Technologies of the self ...

> permit individuals to effect by their own means or with the help of others a certain number of operations on their own bodies and souls, thoughts, conduct, and way of being, so as to transform themselves in order to attain a certain state of happiness, purity, wisdom, perfection, or immortality (Foucault 1988: 18).

Clearly, – and despite the fact that there is nothing especially new or thrilling about Foucault's investigations into this topic – his allowing for the possibility of such "technologies" rules out the possibility of us being completely determined by interdictions and disciplinary mechanisms. To repeat, these Foucauldian ideas on resistance do not amount to more than hints or sketches. However, not only can the sketchiness of these ideas be defended by the observation that resistance cannot usually be predicted and deduced ("In the end, there is no explanation for the man who revolts" (Foucault 1981c: 5).), i.e. that it is unclear what more *can* be said about resistance on a theoretical level, but it should also be clear that even in their very rudimentary form, there ideas are sufficient to show that, according to the Foucauldian theory of power, resistance is not only possible, but even inevitable.

CRITICISM WITHOUT NORMATIVITY?

Commenting on the absence from Foucault's writings of a normative theory for organizing a just society seems imperative for at least two reasons. On the one hand, explaining this Foucauldian notion warrants the vagueness of his distinction between hegemonial and non-hegemonial power. On the other hand, there is hardly any ingredient of Foucault's thought that has been denounced as often as the alleged contradiction between his implicit criticism of modern forms of power and his unwillingness to spell out the normative basis of this criticism.

Nancy Fraser's way of putting this point can be regarded as representative of many other commentators:

> Why is struggle preferable to submission? Why ought domination to be resisted? ... Only with the introduction of normative notions could he [Foucault] begin to tell us what is wrong with the modern power/knowledge regime and why we ought to oppose it (Fraser 1981: 284).

Like his stance on resistance, Foucault's version of a criticism without an elaborated normative theory also seems to me to be much more plausible than his critics allow for. Let me try to explain this somewhat telegraphically, in part by drawing on Foucault's own arguments, in part by submitting some further, non-Foucauldian arguments in his support.

First, I find unconvincing Fraser's insistence that Foucault owes us a normative theory explaining why we should prefer struggle to submission, i.e. why we should resist domination. It seems arguable that the onus of proof does not lie with those who are willing to resist but rather with those who are not. It seems fair to attribute this position to Foucault based on his answer to the question why he is interested in politics: "... why shouldn't I be interested? That is to say, what blindness, what deafness, what density of ideology would have to weigh one down to prevent one from being interested ..." (1974: 168). Perhaps it is also worth adding that in one place Foucault suggests that it is precisely the message of what he calls the humanist tradition that one should not resist the exercises of power. In fact, Foucault even defines humanism as ...

> the totality of discourse through which Western man is told: "Even though you don't exercise power, you can still be a ruler. Better yet, the more you deny yourself the exercise of power, the more you submit to those in power, then the more this increases your sovereignty" (LCMP 221).

Second, one need not be too impressed by the often-heard claim that criticism needs to be based on firm criteria, and that to accept the validity of a criticism is to accept the criteria upon which it is based. After all, there are many good examples in the history of science and philosophy where criticisms were not presented using firm criteria, and many clear cases where the critique was accepted without the criteria.[57] For instance, nothing prevents one from accepting many of Habermas's analyses and criticisms of modern society, while at the same time rejecting his normative utopia of the ideal speech situation.

Third, and this too is an argument that can not be found - as far as I can see - in Foucault's texts, one will usually manage to mobilize much broader popular support for an initiative, say, against new missiles, new nuclear plants, degrading conditions in hospitals, etc., if one abstains from linking this initiative to some global conception of what society should be like. It is enough that we share roughly the same intuitions telling us, e.g. that maltreatment of prisoners or the mentally ill is inhumane - we need not share the same moral or political theory.

Fourth, and here I come to a point that is central for Foucault himself, one might argue that the task of philosophers and intellectuals is not to tell people how they should live their lives, but at most to participate in the struggle for getting rid of vicious forms and mechanisms of power. That is to say, "the intellectual's role is no longer to place himself 'somewhat ahead and to the side' in order to express the stifled truth of the collectivity ..." (LCMP 207-8), his or her task is not to "deduce from the form of what we are what it is impossible for us to do and know" (ENL 46), but rather the task is the more modest one of pointing out that we need not be what we are, that is, showing "the possibility of no longer being, doing, or thinking what we are, do, or think" (ENL 46).

Fifth, one can also advance an argument against the preoccupation with normative theories by reference to the question of priority, and the current state of the left. On the one hand, arguing with Noam Chomsky about political theory in 1974, Foucault held that the criticism of coercive institutions is too important to be left alone until one has developed "an ideal social model" (1974: 171). On the other hand, in 1977, concerned by events in China, Foucault suggested that the progressive groups and movements of the Western world can no longer refer to any socialist country as a model. The left must "call in question the whole important tradition of socialism in a fundamental way - since, after all, everything which this socialist tradition has produced in history must be condemned" (Foucault 1977: 68). Foucault believes that we are thrown back to the year 1830 and that a leftist, progressive theory has to be developed anew from the ground up. Even if one does not agree with this position in all its radicality, its general drift is certainly even more convincing in the 1990's than it was back in 1977.

Sixth, Foucault considers that utopian conceptions of what a just society must look like are obstacles rather than accelerators of social change. ("I think that to imagine another system is to extend our participation in the present one." – (LCMP 230).) For instance, Foucault regards as the central flaw in the development of Soviet society that it tried to implement a conception of "an ultimately liberated human nature" (1974: 174), a conception shaped by the bourgeois societies in which it had been drawn up. More generally, criticizing modern society in terms of justice or human nature is problematic because these are "notions and concepts which have been formed within our culture, within our type of knowledge and our form of philosophy, and that as a result form part of our class system ..." (*ibid.*, 187).

One noteworthy point about this sixth, and perhaps most central, argument is that it expresses a position that has also frequently been attributed to Marx. Several interpreters of Marx have suggested that Marx cannot have wanted to criticize capitalist society for being unjust since justice, "the *juridical* point of view", is but a function of the mode of production. To measure society by the standard of justice is to remain within the superstructure of society and thus to remain reformist rather than become revolutionary.[58]

A further, rather obvious, observation is that the reasoning behind this argument is of course compelling once we accept the thesis of an internal link between power and knowledge. Clearly, there is no reason why power interests, exclusion and manipulation should turn back at the borders of the philosophical field.

Finally, a comment is called for on the alleged contradiction between the above-quoted Foucauldian rejection of the notions of justice and human nature, and the ways these – or similar – notions appear in, or are presupposed by, his own writings. Obviously, Foucault's own interest in the history of modern forms of power is morally motivated, – his own political activities are as good an indication of this as any. He also accepts in the introduction to one of his articles in *Le Nouvel Observateur* the comment that French institutions are "unjust". Moreover, he claims to feel "impatience for liberty"; he speaks out for "bodies and pleasures"; and he calls for "a new form of right", "autonomy", "the right to knowledge" and the creation of "ourselves as a work of art" (ENL 50; HS I 157; P/K 108; PPHC 51, 160; 1983: 236).

That these different pronouncements need not constitute a contradiction or paradox is easily appreciated by returning once more to Rawlsian terminology, to wit, Rawls' distinction between the concept and conceptions of justice, and his distinction between moral intuitions and theories. It seems natural to suggest that Foucault's opposition to justice is not so much an opposition to the *concept* of, and our moral intuitions concerning, justice, as it is an opposition to modern theoretical *conceptions* thereof. After all, what Foucault seeks to establish, in his *historical analyses*, is that political-philosophical rationality either has its roots in pre-modern conceptions of society – and is thus unable even to conceptualize modern power – or that it is intertwined with the rationality that rules the Panoptical society. However, Foucault's criticism of that rationality and the Panoptical society itself need not depend upon these same conceptions of justice or human nature. Most of us that know nothing of modern theories of justice or of humanism still probably feel that there is something fundamentally unjust about the ways prisoners or the mentally ill are treated, or that there is something inhumane about many power mechanisms in science. Drawing, as Foucault does, on these intuitions is not irrational, especially as long as the other factor of the Rawlsian "reflective equilibrium" (1972: 20), i.e. moral theories or "moral geometry" (ibid., 121), stand in doubt. Perhaps most of us also share Foucault's intuition that the construction of new moral or political theories in the future – a possibility implied by Foucault's call for a novel concept of right – has to be constrained by at least the following restriction: "The search for a form of morality acceptable to everybody in the sense that everybody should submit to it strikes me as catastrophic" (PPHC 253–54).

To conclude, let us note that even though the idea used in the last paragraph to make sense of Foucault's drawing on moral intuitions rather than elaborated ethical theories was not formulated by himself, a similar stand is expressed by him with respect to the wider issue of his relation to the Enlightenment tradition. On the one hand, Foucault suggests that we can no longer adopt many of the central doctrines of the Enlightenment, e.g. its humanism, i.e. its assumption of a permanent human nature, its naive opposition between social power and knowledge, or its utopianism. On the other hand, for Foucault, this rejection of central doctrines of the Enlightenment does not imply that we can no longer trust the intuition or attitude that underlies, and emerges from, Kant's writings, and that continues to be at work from Hegel through Nietzsche and Weber up to the Frankfurt School, namely the intuition, ethos or attitude that our historical era calls for a "permanent critique", a questioning of what is given to us as "universal, necessary, obligatory", a readiness to experiment with new forms of social life, and a constant attempt to disconnect the growth of our scientific and technical capabilities from the "intensification of power relations" (ENL 48). Foucault calls this attitude "the critical ontology of ourselves" and explains that ...

> it is to be conceived as an attitude, an ethos, a philosophical life in which the critique of what we are is at one and the same time the historical analysis of the limits that are imposed on us and an experiment with the possibility of going beyond them (ENL 50).

NOTES TO PART II

[1] For example, Robert Dahl and Stephen Lukes write as follows:

> "*A* has power over *B* to the extent that he can get *B* to do something that *B* would not otherwise do" (Dahl 1969: 80).
>
> "... *A* exercises power over *B* when *A* affects *B* in a manner contrary to *B*'s interests" (Lukes 1974: 27).

[2] See Dahl (1969); Harsanyi (1969); Wrong (1979: 12–17).

[3] "A person kept thoroughly confined and supervised, as an individual in a strait-jacket, ... is not a participant in a reciprocal power-relation" (Giddens 1979: 149).

[4] E.g. Wrong (1979); Lukes (1974); Bachrach and Baratz (1969a) and (1969b).

[5] "In general, we understand by 'power' the chance of a man or a number of men to realize their own will in a communal action even against the resistance of others who are participating in the action" (Weber 1970: 80).

[6] Interestingly enough, Lukes's characterization, which tends to push Foucault's conception of power into the camp of (marxist-) structuralist theories of power, can also be countered with the observation that the Althusserian Nicos Poulantzas, whom Lukes considers rightly to be *the* champion of a responsibility-denying conception of power, has in fact claimed that Foucault retains too many ingredients of the Anglosaxon sociological tradition. Little surprise that Poulantzas faults Foucault's genealogy for this closeness (Poulantzas 1978b: 41). In order to defend Foucault against Poulantzas's charge of merely repeating old Anglosaxon ideas, we need to turn to the features of Foucault's theory of power that set him apart from this tradition.

[7] In outlining the Leibnizian distinction below, I follow Rescher (1981: 56–83).

[8] This was already vaguely suggested by Veyne (1981: 63).

[9] The idea of applying the Leibnizian distinction to theories of power has been elaborated most convincingly by Jon Elster, who incidentally attributes the view that power relations are internal to Marx (Elster 1978: 20–27; Elster 1985: 92–95). Elster suggests that we conceive of power as an external relation when we treat power as something that different individuals or groups possess in different degrees, and when we distinguish between groups in society on the basis of how much power they possess. What this kind of approach leaves out, however, is what takes place in the interaction between the groups or individuals.

[10] Foucault's fairly wide concept of power mechanisms allows him to speak of the *exercise* of power even in those cases in which other writers speak of power as a *disposition* rather than an *episode* (or exercise). That is to say, these writers wish to draw a distinction between the cases where threats are actually made, and cases where the power subject submits in the absence of direct interaction (e.g. Wrong 1979: 6). Foucault's claim that power "exists only in action" (P/K 89) seems to blur this distinction. In order to preserve this distinction within the Foucauldian framework, we might suggest that regardless of whether the compliance follows the presentation of threats or not, it is in any case the result of an exercise of power: the two cases (of power as episode or disposition) are distinguishable by the criteria either of whether the power exercise is temporally close or temporally remote from the power subject's action (or non-action), or by which types of different mechanisms of power are involved (e.g. ecological control *versus* threat).

[11] The notion of "ecological control" was suggested by Cartwright (1969: 140).

[12] See, e.g. Kedar (1987).

[13] Lawyers, historians, and political writers like Edward Coke (1552–1634), John Lilburne (1614–1657), Nicolas Boulainvilliers (1688–1749), Charles Hector Estaing (1729–1794), Augustin Thierry (1795–1856), and later biologists, racists and proponents of eugenics (LK).

[14] While the details of this history lie beyond the scope of this study, it is perhaps interesting to note in passing that Foucault's turning around of the Clausewitzian principle is not unique among political philosophers of this century. For instance, in the thirties Antonio Gramsci sought to conceptualize the class struggle for hegemony in terms of the distinction between war of movement and war of position. More precisely, Gramsci analyzed economic crises as bombardments of heavy artillery that have to be employed by the left in order to move towards a frontal attack on the class opponent (Gramsci 1980: 268–71). Another example is of course Carl Schmitt, one of the most eminent reactionary political thinkers of this century, who again in the 1930's defined the essence of politics via the categories of friend and enemy, taking these latter concepts as referring to friend and enemy in a battle of war. Schmitt does not confine this opposition to international politics but applies it to domestic politics as well: "The real possibility of struggle that must always be present in order for us to speak of politics, relates – when we take into account the 'primacy of domestic politics' – not to the war between organized peoples (states or empires), but to civil war" (Schmitt 1932: 20). Schmitt also faulted Clausewitz for having

neglected the importance of a study of war for an understanding of politics (*ibid.*, 21). Surprisingly enough, these parallels between the arch-reactionary Schmitt and Foucault seem not to have been exploited yet by Foucault's interpreters.

[15] For good introductions to Faraday's thought, see Agassi (1971); Berkson (1974); Sachs (1973).

[16] On this notion, see Agassi (1971: 102–16).

[17] See foremost Parfit (1984).

[18] See, for a more detailed discussion, Berkson (1974: 105).

[19] For instance, Jean Baudrillard argues that "if power were this magnetization of the social field then it would hardly any longer meet with resistance" (Baudrillard 1977: 58).

[20] Note also that even aside from the question as to whether or not Foucault himself has made fruitful use of the omnipresence thesis, this idea is in fact the basic starting point of one of the most elaborated theories of power in Anglosaxon sociology. Talcott Parsons' theory of power is based on the premiss that power is "a circulating medium, analogous to money" (Parsons 1969: 256). This medium is spread throughout society, even though it can, at least in part, be "invested" or concentrated. Contrary to a suggestion found in the literature (Kroker 1984), it seems to me that Foucault shares little else with Parsons apart from this idea of the omnipresence and circulating character of power. For example, the very parallel between money and power, and Parsons' attempt to construe political institutions as banks in which we invest our power-money are clearly diametrically opposed to Foucault's anti-economism and his refusal to study power in terms of legitimization. Yet the fact remains that Parsons' theory constitutes a case where the omnipresence assumption finds a non-trivial application.

[21] See, e.g. Anderson and Carlos (1976); Baumgartner, Buckley, Burns and Schuster (1976); Lorrain and White (1971).

[22] *Petit Larousse Illustré* (1988: 320).

[23] Deleuze (1988: 28).

[24] This is an important point to take up, since Foucault has been charged with leaving no room for this possibility. Charles Taylor has accused Foucault of neglecting even a consideration of invisible hand explanation, claiming that Foucault explains coercive institutions exclusively in terms of "Power" as "a strange kind of Schopenhauerian will, ungrounded in human action" (Taylor 1986: 88).

[25] For an analysis of counterfinality phenomena, see Elster (1985).

[26] See Ullmann-Margalit (1978) for a detailed discussion of this distinction.

[27] A fourth, highly interesting argument can be found in Rouse (1987).

[28] Since this idea draws a conclusion from observations by Ian Hacking, a brief look at its original context might soften resistance to the subsequent conclusion.

At a time when the direction of the bulk of basic research in the natural sciences and mathematics is determined by military funding, Hacking (1987) asks us to consider whether it is not natural to assume that the very form of our knowledge has been permanently shaped by this relation. Hacking takes a form of knowledge of a branch of science to be "a structured set of declarative sentences that stand for possibilities, that is, sentences that can be true or false, together with techniques for finding out which ones are true and which ones are false" (*ibid.*, 243). Hacking suggests that as more and more results of weapons research come to figure as exemplars or paradigms of what constitutes good scientific work, it is reasonable to assume that alternative ways of doing research drop out from our screen of possibilities, and weapons research increasingly determines what we regard as scientific knowledge, as meaningful areas of research or as possible candidates for truth and falsehood (*ibid.*, 260).

[29] This point is also stressed by Hacking (1987: 250).

[30] See Lewis (1973) for the analysis of counterfactuals employed here.

[31] Thus, in *L'Ordre du discours*, he formulates the following "methodological demands": (1) "a principle of *reversing*", according to which the seemingly positive figures of author, discipline, or will to truth are to be treated as principles of rarification and exclusion; (2) "a principle of *discontinuity*", according to which discursive practices are to be attended to as often excluding and ignoring one another; (3) "a principle of *specificity*", according to which we are to regard discourses as practices that we impose upon things; and (4) a principle of "*exteriority*", according to which one is to study the external conditions of the possibility of discourses. In the same text Foucault also speaks of four "regulative principles of analyses", these being "the event", "the series", "the regularity", and "condition of possibility". These notions are meant to replace notions of the traditional history of ideas, concepts like creation, unity, originality and meaning, i.e. notions connected to the idea of a subject that freely creates meaning (LD 53–56).

In *Discipline and Punish*, we are again given "four general rules" of analysis: (1) attend to the "possible positive effects" of power and punishment; (2) focus on "techniques" or "tactics" of power use and punishment; (3) do not study "the history of penal law and the history of the human sciences as two separate series"; and (4) investigate how control over the body shapes "the soul" or personal identity (DP 23–24).

In "Two Lectures" (P/K 78–108) the four "methodological precautions" are: (1) study power not on the level of the question of its legitimacy but on the level of how different institutions work on individuals, i.e. "try to locate power at the extreme points of its exercise"; (2) study power in its exercise rather than asking who has power and what are the intentions of the power holder, i.e. study how power shapes subjects; (3) analyze power as "something which circulates", as "a chain" or as a "network"; (4) "conduct an *ascending* analysis of power", and abstain from giving undue importance to ideological effects (P/K 97–102).

Finally, in *History of Sexuality I*, Foucault formulates "four rules to follow", telling us that these are "not intended as methodological imperatives; at most they are cautionary prescriptions": (1) the "rule of immanence", according to which knowledge and power are to be studied as internally related; (2) "rules of continual variations", according to which "relations of power-knowledge are not static forms of distribution, they are 'matrices of transformations'"; (3) the "rule of double conditioning", according to which macrolevel and microlevel power mechanisms are to be regarded as interdependent; and (4) the "rule of the tactical polyvalence of discourses", according to which discourses are to be studied as serving different functions (HS I 98–100).

32 "... genealogy ... must record the singularity of events outside of any monotonous finality ..." (NGH 76).

"An event ... is not a decision, a treaty, a reign, or a battle, but the reversal of a relationship of forces, the usurpation of power, ..." (NGH 88).

Genealogy is thus concerned "to reestablish the various systems of subjection: not the anticipatory power of meaning, but the hazardous play of dominations" (NGH 83).

33 Cf.: "The body is molded by a great many distinct regimes; it is broken down by the rhythms of work, rest, and holidays; it is poisoned by food or values, through eating habits or moral laws; it constructs resistances" (NGH 87).

34 Nietzsche (1968: 481, 552). Here quoted from Nehamas (1985: 42, 173).

35 Cf. Part I of this study, especially Foucault's notion of series.

36 Here and in the following I make use of ideas by G. H. von Wright. See the papers in von Wright (1984).

37 Taking their lead from Collins (1985).

38 Foucault calls this ability of the *dispositif* of the prison to be adaptable to different ends it strategic "saturation" (*remplissement*) (P/K 196).

[39] Linking Elias's and Bakhtin's classical studies with more recent feminist concerns, Dorinda Outram has suggested that Bakhtin's work is to be taken as a corrective of Elias's, and that the important turn from the "carnevalesque" to the *homo clausus* is situated around the French Revolution: "The victory of the *homo clausus* ... was not complete until, through the French Revolution, a representative bourgeois culture of public dominance could be constructed which functioned not only through institutionalized support in the new state, but also ... through the creation of a new political culture defined precisely as that victory of *homo clausus* over both the feminine and the carnivalesque" (Outram 1989: 16). Outram shows convincingly that the French Revolution led to the creation of a political culture from which women were excluded and that this exclusion reflected itself in conceptions of the male and the female body.

[40] The shift from the first to the second is one of the central themes of *DP*.

[41] The latter alternative is of course the central topic of *HS I*.

[42] Cf. Amnesty International (1987: 139).

[43] See e.g. Osnowski (1988).

[44] This feature of Foucault's work is convincingly stressed by Veyne (1981).

[45] *DP*, part two.

[46] *HS I, passim.*

[47] Even though Foucault claims at one point that he is not concerned with interests, I shall ignore this disclaimer and follow his interlocutors in holding that he does in fact provide explanations in terms of interests (P/K 205–6).

[48] There are, of course, many other interesting parallels between Foucauldian genealogy and Latour's work (1987, 1988). See Rouse (1987) for further elaboration on these parallels.

[49] Sociologists of science have only recently attended to this feature of their work. Thus Steve Woolgar has interpreted the irony of the sociological enterprise concerning science as arising from the fact that sociology replaces the account presupposed by the natural scientists (nature determines knowledge) with an account that sees knowledge and nature as determined by scientists' negotiations, quarrels and contingent laboratory settings. Woolgar also distinguishes between "stable" and "dynamic irony": the former is nothing but the replacement of an actor's account with an allegedly more adequate one; the latter signals, additionally, its own transient and provisional character: "dynamic irony asks the reader constantly to recognize the fragility of the ironicist's own account ..." (Woolgar 1983: 260).

[50] Quoted from Baker and Hacker (1983: 4).

[51] See also Hesse (1980: 44–48).

[52] Like MacIntyre (1987) and Taylor (1986).

[53] *Pace* Taylor (1986: 94).

[54] Even though Foucault claims that direct criticism of moral wrongdoing within institutions is not the primary task of genealogical criticism, it is by no means true that he has abstained from criticizing persons and institutions directly. This much has to be firmly said in answer to those many critics who claim that Foucault always depicts power as a "Schopenhauerian will" (Taylor 1986: 88) for which no one is responsible. Foucault's numerous articles and interviews for widely-circulating French daily and weekly papers speak a totally different language. For instance, in an article published in *Le nouvel observateur* in 1972 (Foucault 1972), Foucault attacks the French president, Pompidou, for not having pardoned two men who were subsequently executed with the guillotine. He refers to the executor by name, mentions his hometown and even reveals his salary for carrying out these executions. Foucault also speaks directly of those who are responsible for the prison system and the constant threat of death its inmates are faced with: "the legal administrative body, the Church, the armed police, and, in the shadow, the president of the Republic ... [and] the prison personnel". Furthermore, in the same article Foucault criticizes the prison sentence imposed upon a woman for having had an abortion, and deplores the case of a woman who was imprisoned because she could not pay back a loan. To take another example, in an article for *Libération* published in 1981 (Foucault 1981), Foucault speaks explicitly of "the political class", quotes prisoners' reports on their hunger strike and the unwillingness of the prison warden to negotiate, and he claims that for the ruling political elite the ideals of the French Revolution mean preciously little.

[55] Concerning the possibility of resistance on the basis of Foucault's theory of power, I have already referred to Baudrillard's argument that power *qua* magnetization of the social field simply leaves no room for resistance. Michel Pêcheux has claimed that Foucault's theory of power is marked by a "masked biologism which ... does indeed make revolt strictly unthinkable". Pêcheux believes that on the basis of Foucauldian premisses it is impossible to distinguish between "subjection of human individuals and the procedures of animal domestication" (Pêcheux 1982: 219). Moreover, Nicos Poulantzas has argued that Foucault lacks a subject or agent of resistance, suggesting that only a social class like the proletariat can take this position (Poulantzas 1978a: 137).

[56] The role of this fallacy in sociological and political thought has been studied by Elster (1978).

[57] Agassi (1988b: 213) makes this point in a different context.

[58] On this issue see the papers in Cohen, Nagel and Scanlon (1980). Cf. Lukes (1985).

BIBLIOGRAPHY

1. WRITINGS, EDITIONS, AND INTERVIEWS BY FOUCAULT

1954 *Maladie mentale et personnalité*, P.U.F., Paris.

1960 "Introduction à L'Anthropologie de Kant". Thèse complémentaire pour le doctorat des lettres, Université de Paris, Faculté des lettres et des sciences humaines. This unpublished thesis is preserved at the *Centre Michel Foucault* in Paris.

MC *Madness and Civilization*, translated by J. Barchilon, Pantheon, New York, 1965. Originally published in French as *Folie et déraison. Histoire de la folie à l'âge classique*, Plon, Paris, 1961.

1966 "Une histoire restée muette", *La Quinzaine littéraire*, July 1st.

1968 "Foucault répond à Sartre", *La Quinzaine litteraire*, March 1st-15th.

1969 "Michel Foucault explique son dernier livre", *Magazine littéraire* 28 (April-May): 23-25.

1971a "Entretien avec Michel Foucault", in Bellour 1971: 135-44.

1971b "Foucault responds/2", *Diacritics* vol. 1: 60.

LD *L'ordre du discours*, Gallimard, Paris, 1971.

1972 "Les deux morts de Pompidou", *Le nouvel observateur*, Dec. 4.

BC *The Birth of the Clinic: An Archaeology of Medical Perception* translated by A. Sheridan Smith, Pantheon, New York, 1973. Originally published in French as *Naissance de la clinique. Une archéologie du regard médical*, PUF, Paris, 1963.

OT *The Order of Things. An Archaeology of the Human Sciences*, Tavistock, London, 1974. Originally published in French as *Les mots et les choses. Une archéologie des sciences humaines*, Gallimard, Paris, 1966.

AK *The Archaeology of Knowledge*, translated by A.M. Sheridan Smith, Tavistock, London, 1974. Originally published in French as *L'archéologie du savoir*, Gallimard, Paris, 1969.

1974 (with N. Chomsky), "Human Nature: Justice versus Power", in Elders 1974: 133-98.

1975 *I, Pierre Rivière, having slaughtered my mother, my sister and my brother*, edited by M. Foucault, translated by F. Jellinek, Pantheon, New York. Originally published in French as *Moi, Pierre Rivière, ayant égorgé ma mère, ma soeur et mon frère*, Gallimard, Paris, 1973.

MIP *Mental Illness and Psychology*, translated by A. Sheridan, Harper Colophon, New York.

1976 *Über Strafjustiz, Psychiatrie und Medizin*, translated by W. Seitter, Merve, Berlin.

LCMP *Language, Counter-Memory, Practice: Selected Essays and Interviews*, edited by D.F. Bouchard, Cornell University Press, Ithaca, N.Y., 1977.

1977 "Die Folter, das ist die Vernunft. Ein Gespräch Knut Boesers mit Michel Foucault", *Literaturmagazin*, vol. 8: 60–68.

DP *Discipline and Punish: The Birth of the Prison*, translated by A. Sheridan Smith, Pantheon, New York, 1977. Originally published in French as *Surveiller et punir. Naissance de la prison*. Gallimard, Paris, 1975.

1978 "Politics and the Study of Discourse", *Ideology and Consciousness*, vol. 3: 7–26.

P/K *Power/Knowledge: Selected Interviews & Other Writings 1972–1977*, edited by C. Gordon, Pantheon, New York, 1980.

HS I *The History of Sexuality I: An Introduction*, translated by R. Hurley, Pantheon, New York, 1978. Originally published in French as *Histoire de la sexualité I: La volonté de savoir*, Gallimard, Paris, 1976.

1980 *Herculine Barbin, Being the Recently Discovered Memoirs of a Nineteenth Century French Hermaphrodite*, edited by M. Foucault, translated by R. McDougall, Pantheon, New York. Originally published in French as *Herculine Barbin dite Alexina B.*, Gallimard, Paris, 1978.

1981a "Foucault at the Collège de France II: A Course Summary", *Philosophy and Social Criticism*, vol. VIII: 349–59.

1981b "Il faut tout repenser la lois et la prison", *Libération*, July 6th.

1981c "Is it useless to revolt?", *Philosophy and Social Criticism*, vol. 8: 1–9.

1982 (Edited with A. Farge), *Le désordre des familles*: Lettres de cachet des Archives de la Bastille au XVIIIe siècle, Gallimard, Paris.

1983 "On the Genealogy of Ethics: An Overview of Work in Progress", in Dreyfus and Rabinow 1983: 229–52.

S&P "The Subject and Power", in Dreyfus and Rabinow 1983: 208–26.

HS II *The Use of Pleasure*, translated by R. Hurley, Pantheon, New York, 1985. Originally published in French as *Histoire de la sexualité 2: L'usage de plaisirs*, Gallimard, Paris, 1984.

HS III *The Care of the Self*, translated by R. Hurley, Pantheon, New York, 1985. Originally published in French as *Histoire de la sexualité 3: Le souci de soi*, Gallimard, Paris, 1984.

FR *The Foucault Reader*, edited by P. Rabinow, Penguin Books, Harmondsworth, 1984.

ENL "What is Enlightenment?", in *FR* 32–50.

NGH "Nietzsche, Genealogy, History", in *FR* 76–100.

LK *Vom Licht des Krieges zur Geburt der Geschichte*, translated by W. Seitter, Merve, Berlin, 1986.

QM "Questions of Method: An Interview with Michel Foucault", in Baynes *et al.* 1987: 100–17.

1987 (and B. Blanchot), *Foucault. Blanchot*, Zone Books, New York.

1988 "Technologies of the Self", in Martin *et al.* 1988: 16–49.

PPHC *Politics. Philosophy. Culture. Interviews and Other Writings 1977–1984*, edited by L.D. Kritzman, Routledge, New York and London, 1988.

DI "The Dangerous Individual", in *PPHC* 125–51.

2. OTHER SOURCES

Ackermann, R. (1989), "The New Experimentalism", *British Journal for the Philosophy of Science*, vol. 40: 185–90.

Agassi, J. (1963), *Towards an Historiography of Science, History and Theory*, Beiheft 2.

Agassi, J. (1971), *Faraday as a Natural Philosopher*, The University of Chicago Press, Chicago and London.

Agassi, J. (1988a), "Cohen contra Kuhn – clues hidden in many details", in Agassi 1988b: 123–30.

Agassi, J. (1988b), *The Gentle Art of Philosophical Polemics*, Open Court, La Salle, Ill..

Althusser, L. (1970a), "From *Capital* to Marx's Philosophy", in Althusser and Balibar 1970: 11–69.

Althusser, L. (1970b), "The Errors of Classical Economics: Outline of a Concept of Historical Time", in Althusser and Balibar 1970: 91–118.

Althusser, L. and E. Balibar (1970), *Reading Capital*, translated by B. Brewster, NLB, London. (The French original was published in 1968.)

Althusser, L. (1977), *For Marx*, translated by B. Brewster, Verso, London. (The French original was published in 1965.)

Amnesty International (1987), *United States of America. The Death Penalty*,

Amnesty International Publications, London.

Anderson, B. and M.L. Carlos (1976), "What is social network theory?", in Burns and Buckley 1976: 27–51.

Bachelard, G. (1927), *Essai sur la connaissance approchée*, Vrin, Paris.

Bachelard, G. (1940), *La philosophie du Non*, P.U.F., Paris.

Bachelard, G. (1951), *L'activité rationaliste de la physique contemporaine*, P.U.F., Paris.

Bachelard, G. (1964), *The Psychoanalysis of Fire*, translated by A.C. Ross, Routledge & Kegan Paul, London. (The French original was published in 1938.)

Bachelard, G. (1971), *Epistemologie. Ausgewählte Texte*, edited by D. Lecourt, Ullstein, Frankfurt am Main.

Bachelard, G. (1978), *Die Philosophie des Nein*, translated by G. Schmidt and M. Tietz, B. Heymann, Wiesbaden.

Bachelard, G. (1984), *The New Scientific Spirit*, translated by A. Goldhammer, Beacon Press, Boston. (The French original was published in 1934.)

Bachrach, P. and M.S. Baratz (1969a), "Decisions and Non-decisions: An Analytical Framework", in Bell *et al.* 1969: 100–109.

Bachrach, P. and M.S. Baratz (1969), "Two faces of power", in Bell *et al.* 1969: 94–99.

Baker, P.M.S. and G.P. Hacker (1983), *An Analytical Commentary on Wittgenstein's Philosophical Investigations*, Basil Blackwell, Oxford.

Bakhtin, M. (1968), *Rabelais and His World*, translated by H. Iswolsky, Harvard University Press, Cambridge, Mass. (The Russian original was first published in 1965.)

Barnes, B. (1977), *Interests and the Growth of Knowledge*, Routledge and Kegan Paul, London.

Barnes, B., *The Nature of Power* (1988), Polity Press, Cambridge.

Barnes, B. and S. Shapin (eds.) (1979), *Natural Order. Historical Studies of Scientific Culture*, SAGE, London.

Baudrillard, J. (1977), *Oublier Foucault*, éditions galilée, Paris.

Baumgartner, T., W. Buckley, T.R. Burns and P. Schuster (1976), "Metapower and the structuring of social hierarchies", in Burns and Buckley 1976: 215–88.

Baynes, K., J. Bohman, and T. McCarthy (eds.) (1987), *After Philosophy. End or Transformation?*, Massachusetts Institute of Technology, Cambridge, Mass.

Bell, R., D.V. Edwards and & R.H. Wagner (eds.) (1969), *Political Power. A Reader in Theory and Research*, The Free Press, New York,

Collier-Macmillan Ltd., London.

Bellour, R. (1971), *Le Livre des autres*, Editions de l'Herne, Paris.

Benton, T. (1984), *The Rise and Fall of Structuralist Marxism. Althusser and his Influence*, MacMillan, Houndsmills, Basingstoke.

Berkson, W. (1974), *Fields of Force. The Development of a World View from Faraday to Einstein*, Routledge and Kegan, London.

Blanchot, M. (1987), "Michel Foucault as I Imagine Him", translated by J. Mehlman, in Foucault and Blanchot 1987: 61–109.

Bloch, M. (1924), *Les rois thaumaturges*, Paris.

Bloch, M. (1953), *The historian's craft*, translated by P. Putnam, Vintage Books, New York. (The French original was first published in 1941.)

Bloch, M. (1962), *Feudal Society*, translated by L.A. Manyon, Routledge & Kegan Paul, London.

Bloor, D. (1976), *Knowledge and Social Imagery*, Routledge and Kegan Paul, London.

Boucher, D. (1985), *Texts in Context. Revisionist Methods for Studying the History of Ideas*, Martinus Nijhoff, Dordrecht.

Bouligand, G. *et al.* (eds.) (1959), *Hommage à Gaston Bachelard*, P.U.F., Paris.

Bourdé G. and H. Martin (1983), *Les écoles historiques*, Éditions du Seuil, Paris.

Bourdieu, P. (1988), *Homo Academicus*, translated by P. Collier, Polity Press, Cambridge. (The French original was published in 1984.)

Braudel, F. (1972–74), *The Mediterranean and the Mediterranean World in the Age of Philip II*, translated by S. Reynolds, 2 vols., Harper and Row, New York. (The French original was first published in 1949.)

Braudel, F. (1980), *On history*, translated by S. Matthews, The University of Chicago Press, Chicago.

Brühmann, H. (1980), *'Der Begriff des Hundes bellt nicht'. Das Objekt der Geschichte der Wissenschaften bei Bachelard und Althusser*, B. Heymann, Wiesbaden.

Burguière, A. (1982), "The Fate of the History of *Mentalités* in the *Annales*", *Comparative Studies in Society and History*, vol. 24: 424–37.

Burguière, A. (ed.) (1986), *Dictionnaire des sciences historiques*, P.U.F., Paris.

Burke, P. (1986), "Strength and Weaknesses of the History of Mentalities", *History of European Ideas*, vol. 7: 439–51.

Burns, T. and W. Buckley (eds.) (1976), *Power and Control. Social Structures and Their Transformations*, SAGE Studies in International Sociology 6, SAGE Publications, London and Beverly Hills, Cal.

Canguilhem, G. (1955), *La formation du concept de réflexe aux XVIIe et XVIIIe siècles*, P.U.F., Paris.

Canguilhem, G. (1959), "Sur une épistémologie concordataire", in Bouligand *et al.* 1959: 3–12.

Canguilhem, G. (1968), *Études d'histoire et de philosophie des sciences*, Vrin, Paris.

Canguilhem, G. (1978), *On the Normal and the Pathological*, translated by C.R. Fawcett, Reidel, Dordrecht. (The French original was first published in 1943.)

Canguilhem, G. (1988), *Ideology and Rationality in the History of the Life Sciences*, translated by A. Goldhammer, The MIT Press, Cambridge, Mass. (The French original was first published in 1977.)

Cartwright, D. (1969), "Influence, Leadership, Control", in Bell *et al.* 1969: 123–65.

Cartwright, N. (1983), *How the Laws of Physics Lie*, Oxford University Press, Oxford.

Cassirer, E. (1932), *Die Philosophie der Aufklärung*, second edition, Mohr, Tübingen.

Cavaillès, J. (1987), *Sur la logique et la théorie de la science*, Librairie Philosophique J. Vrin, fourth edition, Paris.

Chartier, R. (1982), "Intellectual History or Sociocultural History? The French Trajectories", in LaCapra and Kaplan 1982: 13–46.

Chaunu, P. (1970), "L'histoire sérielle. Bilan et perspectives", *Revue historique*, vol. 494: 297–320.

Childe, V.G. (1956), *Piecing Together the Past. The Interpretation of Archaeological Data*, Routledge and Kegan Paul, London.

Childe, V.G. (1962), *A Short Introduction to Archaeology*, Collier Books, New York.

Clausewitz, K. von (1918), *Vom Kriege*, B. Behr's Verlag, 13th edition, Berlin und Leipzig.

Clegg, S. (1979), *The Theory of Power and Organization*, Routledge and Kegan Paul, London.

Cohen, I.B. (1980), *The Newtonian Revolution*, Cambridge University Press, Cambridge.

Cohen, I.B. (1985), *Revolution in Science*, Harvard University Press, Cambridge, Mass.

Cohen, M., T. Nagel, and T. Scanlon (eds.) (1980), *Marx, Justice, and History*, Princeton University Press, Princeton.

Collins, H.M. (1982), "Special Relativism", *Social Studies of Science*, vol. 12: 139–43.

Collins, H.M. (1985), *Changing Order: Replication and Induction in Scientific Practice*, SAGE, London.
Condren, C. (1985), *The Status and Appraisal of Classical Texts. An Essay on Political Theory, Its Inheritance, and the History of Ideas*, Princeton University Press, Princeton.
Cousins M. and A. Hussain (1984), *Michel Foucault*, MacMillan, Houndsmills.
Crombie, A.C. (1981), "Philosophical Presuppositions and Shifting Interpretations of Galileo", in Hintikka et al. 1981: 271–86.
Crombie, A.C. (1986), "What is the history of science?", *History of European Ideas*, vol. 7: 21–31.
Crombie, A.C. (1988), "Designed in the Mind: Western Visions of Scien ce, Nature and Humankind", *History of Science*, vol. 26: 1–12.
Dahl, R.A. (1969), "The Concept of Power", in Bell *et al.* 1969: 79-93.
Deleuze, G. (1988), *Foucault*, translated by S. Hand, The Athlone Press, London. (The French original was published in 1986.)
Dreyfus, H. and P. Rabinow (1983), *Beyond Structuralism and Hermeneutics*, The Harvester Press, second edition, Brighton.
Duhem, P. (1905), *Les origines de la statique*, vol. 1, A. Herrmann, Paris.
Duhem, P. (1906a), *Etudes sur Léonard de Vinci*, vol. 1, A. Herrmann, Paris.
Duhem, P. (1906b), *La Théorie physique. Son objet et sa structure*, Chevalier & Rivière, Paris.
Duhem, P. (1914), *Le Système du monde. Histoire des doctrines cosmologiques de Platon à Copernic*, A. Herrmann, Paris.
Duhem, P. (1969), *To save the phenomena. An essay on the idea of physical theory from Plato to Galileo*, translated by E. Doland and C. Maschler, The University of Chicago Press, Chicago.
Dunn, J. (1968), "The Identity of the History of Ideas", *Philosophy*, vol. 43: 85–104.
Elders, F. (ed.) (1974), *Reflexive Water. The Basic Concerns of Mankind*, Condor, London.
Elias, N. (1978), *Über den Prozeß der Zivilisation. Soziogenetische und psychogenetische Untersuchungen*, Suhrkamp, Frankfurt am Main. (The German original was first published in 1939.)
Elkana, Y. (1981), "A Programmatic Attempt at an Anthropology of Knowledge", in Mendelsohn and Elkana 1981: 1–76.
Elster, J. (1978), *Logic and Society. Contradictions and Possible Worlds*, John Wiley & Sons, Chichester.
Elster, J. (1985), *Making Sense of Marx*, Cambridge University Press,

Cambridge.

Eribon, D. (1989), *Michel Foucault*, Flammarion, Paris.

Febvre, L. (1955), "Sur Einstein et sur l'histoire", *Annales E.S.C.*, vol. 10: 305–12.

Febvre, L. (1982), *The Problem of Unbelief in the Sixteenth Century. The Religion of Rabelais*, translated by B. Gottlieb, Harvard University Press, Cambridge, Mass. (The French original was first published in 1942.)

Fekete, J. (ed.) (1984), *The Structural Allegory. Reconstructive Encounters with the French Thought*, University of Minnesota Press, Minneapolis.

Feyerabend, P. (1984), *Wissenschaft als Kunst*, Suhrkamp, Frankfurt am Main.

Feyerabend, P. (1987), *Farewell to Reason*, Verso, London and New York.

Fichant, M. (1969), "L'idée d'une histoire des science", in Pêcheux and Fichant 1969: 49–139.

Fleck, L. (1980), *Entstehung und Entwicklung einer wissenschaftlichen Tatsache. Einführung in die Lehre vom Denkstil und Denkkollektiv*, edited and introduced by L. Schäfer and T. Schnelle, Suhrkamp, Frankfurt am Main. (First published in 1935.)

Fraser, N. (1981), "Foucault on Modern Power: Empirical Insights and Normative Confusions", *Praxis International*, vol. 1: 272–87.

Fraser, N. (1985), "Michel Foucault: A 'Young Conservative'?", *Ethics*, vol. 96: 165–84.

Furet, F. (1985), "Quantitative methods in history", translated by D. Denby, in Le Goff and Nora 1985: 12–27.

Galison, P. (1987), *How Experiments End*, The University of Chicago Press, Chicago.

Galison, P. (1988), "History, Philosophy, and the Central Metaphor", *Science in Context*, vol. 2: 197–212.

Giddens, A. (1979), *Central Problems in Social Theory. Action, Structure and Contradiction in Social Analysis*, The Macmillan Press, London and Basingstoke.

Giere, R.N. (1988), *Explaining Science. A Cognitive Approach*, The University of Chicago Press, Chicago and London.

Goldmann, L. (1969), *The Human Sciences and Philosophy*, translated by H.V. White and R. Anchor, Cape, London. (The French original was first published in 1952.)

Gramsci, A. (1980), *Zu Politik, Geschichte und Kultur*, edited by G. Zamis, Röderberg, Frankfurt am Main.

Greenleaf, W.F. (1966), *Oakeshott's Philosophical Politics*, Longmans,

London.

Gutting, G. (1989), *Michel Foucault's archaeology of scientific reason*, Cambridge University Press, Cambridge.

Hacking, I. (1975), *The Emergence of Probability. A Philosophical Study of Early Ideas about Probability, Induction and Statistical Inference*, Cambridge University Press, Cambridge.

Hacking, I. (1982), "Language, Truth, and Reason", in Hollis and Lukes 1982: 48–66.

Hacking, I. (1983a), "The Accumulation of Styles of Scientific Reasoning", in Henrich 1983: 453–65.

Hacking, I. (1983b), *Representing and intervening. Introductory topics in the philosophy of natural science*, Cambridge University Press, Cambridge.

Hacking, I. (1984), "Five parables", in Rorty *et al.* 1984: 103–24.

Hacking, I. (1985), "Styles of Scientific Reasoning", in Rajchman and West 1985: 145–65.

Hacking, I. (1987), "Weapons Research and the Form of Scientific Knowledge", *Canadian Journal of Philosophy*, Supplementary Volume 12: 237–61.

Hacking, I. (1988), "The Participant Irrealist At Large in the Laboratory", *British Journal for the Philosophy of Science*, vol. 39: 277–94.

Hakosalo, H. (1991), *Bio-Power and Pathology: Science and Power in the Foucauldian Histories of Medicine, Psychiatry and Sexuality*, Reports from the Department of History, University of Oulu (Finland).

Harris, E.C. (1979), *Principles of Archaeological Stratigraphy*, Academic Press, London.

Harsanyi, J.C. (1969), "Measurement of Social Power, Opportunity Costs, and the Theory of Two-person Bargaining Games", in Bell *et al.* 1969: 226–38.

Henrich, D. (ed.) (1983), *Kant oder Hegel*, Klett-Cotta, Stuttgart.

Hesse, M. (1980), *Revolutions and Reconstructions in the Philosophy of Science*, Harvester, Brighton.

Hintikka, J., C.D. Gruender, E. Agazzi (eds.) (1981), *Theory Change, Ancient Axiomatics, and Galileo's Methodology*, Reidel, Dordrecht.

Hirsh, A. (1981), *The French New Left: An Intellectual History from Sartre to Gorz*, South End Press, Boston.

Hollis, M. and S. Lukes (eds.) (1982), *Rationality and Relativism*, Blackwell, Oxford.

Holton, G. (1988), *Thematic Origins of Scientific Thought. Kepler to Einstein*, Harvard University Press, Cambridge, Mass. (First published in 1973.)

Hoy, D.C. (ed.) (1986), *Foucault: A Critical Reader*, Basil Blackwell, New York.

Hughes G.E. and M.J. Cresswell (1968), *An Introduction to Modal Logic*, Methuen, London.

Huizinga, J. (1970), *The Waning of the Middle Ages. A Study of the Forms of Life, Thought and Art in France and the Netherlands in the XIVth and XVth Centuries*, translated by F. Hopman, Arnold, London. (First published in 1924.)

Jacob, F. (1974), *The Logic of Living Systems. A History of Heredity*, translated by B.E. Spillmann, Allan Lane, London. (The French original was published in 1970.)

Kedar, L. (ed.) (1987), *Power through discourse*, Ablex, Norwood, N.J.

Knorr-Cetina, K. (1981), *The Manufacture of Knowledge. An Essay on the Constructivist and Contextual Nature of Science*, Pergamon, Oxford.

Knorr-Cetina, K. (1983), "The Ethnographic Study of Scientific Work: Towards a Constructivist Interpretation of Science", Knorr-Cetina and Mulkay 1983: 115–40.

Knorr-Cetina, K. and M. Mulkay (eds.) (1983), *Science Observed*, SAGE Publication, London and Beverly Hills.

Koyré, A. (1957), *From the Closed World to the Infinite Universe*, The Johns Hopkins University Press, Baltimore and London.

Koyré, A. (1973), *Études d'histoire de la pensée scientifique*, Gallimard, Paris.

Koyré, A. (1978), *Galileo Studies*, translated by J. Mepham, The Harvester Press, Hassocks. (The French original was first published in 1939.)

Kragh, H. (1987), *An Introduction to the Historiography of Science*, Cambridge University Press, Cambridge.

Kroker, A. (1984), "Modern Power in Reverse Image: The Paradigm Shift of Michel Foucault and Talcott Parsons", in Fekete 1984: 74–103.

Kuhn, T. (1970), *The Structure of Scientific Revolutions*, second edition, The University of Chicago Press, Chicago.

Kusch, M. (1989), *Language as Calculus vs. Language as Universal Medium. A Study in Husserl, Heidegger and Gadamer*, Kluwer, Dordrecht.

Körner, S. (1984), *Metaphysics: its structure and function*, Cambridge University Press, Cambridge.

LaCapra, D. and St.L. Kaplan (eds.) (1982), *Modern European Intellectual History. Reappraisals and New Perspectives*, Cornell University Press, Ithaca and London.

Lakatos, I. (1974), "Falsification and the Methodology of Scientific Research Programmes", in Lakatos and Musgrave 1974: 91–196.

Lakatos, I. and A. Musgrave (eds.) (1974), *Criticism and the Growth of Knowledge*, Cambridge University Press, Cambridge.

Lakatos, I. (1978), *The methodology of scientific research programmes*, Philosophical Papers Volume 1, edited by J. Worrall and G. Currie, Cambridge University Press, Cambridge.

Latour, B. (1983), "Give me a Laboratory and I will Raise the World", in Knorr-Cetina and Mulkay 1983: 141–70.

Latour, B. (1987), *Science in Action*, Open University Press, Milton Keynes.

Latour, B. (1988), *The Pasteurization of France*, Harvard University Press, Cambridge, Mass.

Latour, B. and S. Woolgar (1986), *Laboratory Life. The Construction of Scientific Facts*, Introduction by J. Salk, with a new postscript by the authors, Princeton University Press, Princeton, N.J. (First published in 1979.)

Laudan, L. (1977), *Progress and Its Problems. Towards a Theory of Scientific Growth*, Routledge and Kegan Paul, London and Henley.

Laudan, L. (1984), *Science and Values. The Aims of Science and Their Role in Scientific Debate*, University of California Press, Berkeley.

Laudan, L., A. Donovan, R. Laudan, P. Barker, H. Brown, J. Leplin, P. Thagard, S. Wykstra (1986), "Scientific Change: Philosophical Models and Historical Research", *Synthese*, vol. 69: 141–223.

Lawrance, C. (1980), "The Nervous System and Society in the Scottish Enlightenment", in Rousseau and Porter 1980: 19–40.

Le Roy Ladurie, E. (1972), *Times of Feast, Times of Famine: A History of the Climate since the Year 1000*, translated by B. Bray, George Allen & Unwin, London. (The French original was published in 1969.)

Le Roy Ladurie, E. (1981), *The Mind and Method of the Historian*, translated by S. Reynolds and B. Reynolds, The University of Chicago Press, Chicago. (The French original was published in 1978.)

Lecourt, D. (1975), *Marxism and Epistemology. Bachelard, Canguilhem and Foucault*, translated by B. Brewster, NLB, London. (The book contains two parts, "Part one: Gaston Bachelard's historical epistemology" (pp. 23–118, first published in French in 1969), and "Part two: For a critique of epistemology" (pp. 119–216, first published in French in 1972).)

Le Goff, J. (1985), "Mentalities: a history of ambiguities", translated by D. Denby, in Le Goff and Nora 1985: 166–80.

Le Goff, J. and P. Nora (eds.) (1985), *Constructing the past. Essays in historical methodology*, Cambridge University Press, Cambridge.

Lemert, C. and G. Gillan (1982), *Michel Foucault. Social Theory as Transgression*, Columbia University Press, New York.

Lepenies, W. (1978), "Wissenschaftsgeschichte und Disziplingeschichte", *Geschichte und Gesellschaft*, vol. 4: 437–51.

Leslie, M. (1970), "In defense of Anachronism", *Political Studies*, vol. XVIII: 433–47.

Lévi-Strauss, C. (1966), *The Savage Mind*, Weidenfold and Nicolson, London. (The French original was published in 1962.)

Lévi-Strauss, C. (1967), *Structural Anthropology*, translated by C. Jacobson and B.G. Schoepf, Anchor, Garden City, N.Y. (The French original was published in 1958.)

Lévi-Strauss, C. (1976), *Tristes Tropiques*, translated by J. and D. Weightman, Penguin Books, New York. (The French original was published in 1955.)

Lewis, D. (1973), *Counterfactuals*, Harvard University Press, Cambridge, Mass.

Lewis, D. (1983a), "Counterpart Theory and Quantified Modal Logic", in Lewis 1983b: 26–46.

Lewis, D. (1983b), *Philosophical Papers*, Vol. I, Oxford University Press, Oxford.

Lorrain, F. and H. C. White (1971), "Structural Equivalence of Individuals in Social Networks", *Journal of Mathematical Sociology*, vol. 1: 49–80.

Lovejoy, A. (1960), *Essays in the History of Ideas*, Capricorn Books, New York. (First published in 1948.)

Lovejoy, A. (1973), *The Great Chain of Being. A Study of the History of an Idea*, eleventh printing, Harvard University Press, Cambridge, Mass. (First published in 1936.)

Lukes, S. (1974), *Power: A Radical View*, Macmillan, Houndmills.

Lukes, S. (1985), *Marxism and Morality*, Clarendon Press, Oxford.

Lukes, S. (1986a), "Introduction", in Lukes 1986b: 1–18.

Lukes, S. (ed.) (1986b), *Power*, Blackwell, Oxford.

MacIntyre, A. (1987), "Relativism, Power, and Philosophy", in Baynes *et al.* 1987: 385–411.

MacIntyre, A. (1988), *Whose Justice? Which Rationality?*, Duckworth, London.

Mandrou, R. (1975), *Introduction to modern France 1500–1640. An essay in historical psychology*, translated by R.E. Hallmark, Holmes & Meier, New York. (The French original was published in 1961.)

Martin, L.H., H. Gutman & P.H. Hutton (eds.) (1988), *Technologies of the Self. A Seminar with Michel Foucault*, Tavistock, London.
Mendelsohn, E. and Y. Elkana (eds.) (1981), *Sciences and Cultures. Anthropological and Historical Studies of the Sciences*, Reidel, Dordrecht.
Merquior, J.G. (1985), *Foucault*, Montana, London.
Merquior, J.G. (1986), *From Prague to Paris. A Critique of Structuralist and Post-Structuralist Thought*, Verso, London.
Minogue, K. (1959), "Power in Politics", *Political Studies*, vol. 7: 269–83.
Nehamas, A. (1985), *Nietzsche. Life as Literature*, Harvard University Press, Cambridge, Mass.
Nietzsche, F. (1968), *The Will to Power*, translated by W. Kaufmann and R.J. Hollingdale, Vintage Press, New York.
O'Farrell, C. (1989), *Foucault. Historian or Philosopher*, MacMillan, Houndsmills.
Osnowski, R. (ed.) (1988), *Menschenversuche*, Kölner Volksblatt Verlag, Köln.
Outram, D. (1989), *The Body and the French Revolution. Sex, Class, and Political Culture*, Yale University Press, New Haven and London.
Parfit, D. (1984), *Reasons and Persons*, Clarendon Press, Oxford.
Parsons, T. (1969), "On the Concept of Political Power", in Bell *et al.* 1969: 251–84.
Pêcheux, M. (1969), "Idéologie et histoire des sciences. Les effets de la coupure galiléenne en physique et en biologie", in Pêcheux and Fichant 1969: 13–47.
Pêcheux, M. and M. Fichant (1969), *Sur l'histoire des sciences*, François Masperio, Paris.
Pêcheux, M. (1982), *Language, Semantics and Ideology*, translated by H. Nagpal, Macmillan, London. (The French original was published in 1975.)
Petit Larousse Illustré 1988, Larousse, Paris.
Pocock, J.G.A. (1972), *Politics, Language and Time*, Methuen, London.
Popper, K. (1974), "Normal Science and its Dangers", in Lakatos and Musgrave 1974: 51–58.
Poulantzas, N. (1978a), *Political Power and Social Classes*, translated by T. O'Hagan, Verso, London. (The French original was published in 1968.)
Poulantzas, N. (1978b), *Staatstheorie. Politischer Überbau, Ideologie, Sozialistische Demokratie*, translated by H. Arenz, T. Brackmann, H. Friedhoff and R. Löper, VSA, Hamburg. (The French original was

published in 1977.)
Rajchman, J. (1984), *Michel Foucault: the freedom of philosophy*, Columbia University Press, New York.
Rajchman, J. and C. West (eds.) (1985), *Post-Analytic Philosophy*, Columbia University Press, New York.
Rawls, J. (1972), *A Theory of Justice*, Clarendon Press, Oxford.
Reingold, N. (1980), "Through Paradigm-Land to a Normal History of Science", *Social Studies of Science*, vol. 10: 475–96.
Rescher, N. (1981), *Leibniz's Metaphysics of Nature*, Reidel, Dordrecht.
Revel, J. (1986a), "Foucault, Michel, 1926–1984", in Burguière 1986: 290–92.
Revel, J. (1986b), "Mentalités", in Burguière 1986: 451.
Rorty, R., J.B. Schneewind and Q. Skinner (eds.) (1984), *Philosophy in History*, Cambridge University Press, Cambridge.
Rouse, J. (1987), *Knowledge and Power. Toward a Political Philosophy of Science*, Cornell University Press, Ithaca and London.
Rousseau, G.S. and R. Porter (eds.) (1980), *The Ferment of Knowledge. Studies in the historiography of eighteenth-century science*, Cambridge University Press, Cambridge.
Sachs, M. (1973), *The Field Concept in Contemporary Science*, Charles C. Thomas, Springfield, Ill.
Sarton, G. (1975), *Introduction to the History of Science. Volume I: From Homer to Omar Khayyam*, Krieger, Huntington, N.Y.
Sartre, J.-P. (1969), *Being and Nothingness*, translated by H. Barnes, Methuen & Co., London. (The French original was published in 1943.)
Schmitt, K. (1922), *Der Begriff des Politischen*, Duncker & Humblot, München und Leipzig.
Shapin, S. (1982), "History of Science and Its Sociological Reconstruction", *History of Science*, vol. 20: 157–211.
Shapin S. and S. Schaffer (1985), *Leviathan and the Air-Pump. Hobbes, Boyle, and the Experimental Life*, Princeton University Press, Princeton.
Shapin, S. (1988), "Robert Boyle and Mathematics: Reality, Representation, and Experimental Practice", *Science in Context*, vol. 2: 23–58.
Sheridan, A. (1980), *Michel Foucault. The Will to Truth*, Tavistock Publications, London.
Siegel, H. (1987), *Relativism Refuted*, Reidel, Dordrecht.
Skinner, Q. (1969), "Meaning and Understanding in the History of Ideas", *History and Theory*, vol. 8: 3–53.
Smart, B. (1985), *Michel Foucault*, Tavistock, London.
Steiner, G. (1971), "The mandarin of the hour - Michel Foucault", *New*

York Review of Books, February 28: 8 & 28–31.

Stoianovich, T. (1976), *French Historical Method. The Annales Paradigm*, Cornell University Press, Ithaca and London.

Taylor, C. (1986), "Foucault on Freedom and Truth", in Hoy 1986: 69–102.

Toulmin, S. (1972), *Human Understanding, Vol. I. General Introduction and Part I*, Clarendon Press, Oxford.

Turner, S. (1989), "Depolicizing Power", *Social Studies of Science*, vol. 19: 533–60.

Ullmann-Margalit, E. (1978), "Invisible-Hand Explanations", *Synthese*, vol. 39: 263–91.

Veyne, P. (1981), *Der Eisberg der Geschichte. Foucault revolutioniert die Historie*, translated by K. Tholen-Struthoff, Berlin. (The French original was published in 1979.)

Veyne, P. (1984), *Writing History. Essay on Epistemology*, translated by M. Moore-Rinvolucri, Manchester University Press, Manchester. (The French original was published in 1971.)

Vilar, P. (1976), *A History of Gold and Money 1450–1920*, translated by J. White, NLB, London. (The Spanish original was published in 1969.)

Vilar, P. (1985), "Constructing Marxist history", translated by I. Patterson, in Le Goff and Nora 1985: 47–80.

Waldenfels, B. (1983), *Phänomenologie in Frankreich*, Suhrkamp, Frankfurt am Main.

Weber, M. (1970), *From Max Weber*, translated, edited and with an Introduction by H.H. Gerth and C. Wright Mills, Routledge and Kegan Paul, London.

Wilson, E.O. (1977), "Biology and the Social Sciences", *Daedalus*, vol. 2: 127–40.

Wisan, W.L. (1981), "Galileo and the emergence of a new scientific style", in Hintikka *et al.* 1981: 311–39.

Woolgar, S. (1983), "Irony in the Social Sciences", in Knorr-Cetina and Mulkay 1983: 239–66.

Woolgar, S. (ed.) (1988), *Knowledge and Reflexivity. New Frontiers in the Sociology of Knowledge*, SAGE, London.

Wootton, D. (1988), "Lucien Febvre and the Problem of Unbelief in the Early Modern Period", *Journal of Modern History*, vol. 60: 695–730.

Wright, G.H. von (1984), *Truth, Knowledge, and Modality*, Blackwell, Oxford.

Wrong, D.H. (1979), *Power. Its Forms, Bases and Uses*, Blackwell, Oxford.

INDEX OF NAMES

INDEX OF SUBJECTS

SYNTHESE LIBRARY

Studies in Epistemology, Logic, Methodology, and Philosophy of Science

1. J. M. Bochénski, *A Precis of Mathematical Logic*. Translated from French and German by O. Bird. 1959 ISBN 90-277-0073-7
2. P. Guiraud, *Problèmes et méthodes de la statistique linguistique*. 1959 ISBN 90-277-0025-7
3. H. Freudenthal (ed.), *The Concept and the Role of the Model in Mathematics and Natural and Social Sciences*. 1961 ISBN 90-277-0017-6
4. E. W. Beth, *Formal Methods*. An Introduction to Symbolic Logic and to the Study of Effective Operations in Arithmetic and Logic. 1962 ISBN 90-277-0069-9
5. B. H. Kazemier and D. Vuysje (eds.), *Logic and Language*. Studies dedicated to Professor Rudolf Carnap on the Occasion of His 70th Birthday. 1962 ISBN 90-277-0019-2
6. M. W. Wartofsky (ed.), *Proceedings of the Boston Colloquium for the Philosophy of Science, 1961–1962*. [Boston Studies in the Philosophy of Science, Vol. I] 1963 ISBN 90-277-0021-4
7. A. A. Zinov'ev, *Philosophical Problems of Many-valued Logic*. A revised edition, edited and translated (from Russian) by G. Küng and D.D. Comey. 1963 ISBN 90-277-0091-5
8. G. Gurvitch, *The Spectrum of Social Time*. Translated from French and edited by M. Korenbaum and P. Bosserman. 1964 ISBN 90-277-0006-0
9. P. Lorenzen, *Formal Logic*. Translated from German by F.J. Crosson. 1965 ISBN 90-277-0080-X
10. R. S. Cohen and M. W. Wartofsky (eds.), *Proceedings of the Boston Colloquium for the Philosophy of Science, 1962–1964*. In Honor of Philipp Frank. [Boston Studies in the Philosophy of Science, Vol. II] 1965 ISBN 90-277-9004-0
11. E. W. Beth, *Mathematical Thought*. An Introduction to the Philosophy of Mathematics. 1965 ISBN 90-277-0070-2
12. E. W. Beth and J. Piaget, *Mathematical Epistemology and Psychology*. Translated from French by W. Mays. 1966 ISBN 90-277-0071-0
13. G. Küng, *Ontology and the Logistic Analysis of Language*. An Enquiry into the Contemporary Views on Universals. Revised ed., translated from German. 1967 ISBN 90-277-0028-1
14. R. S. Cohen and M. W. Wartofsky (eds.), *Proceedings of the Boston Colloquium for the Philosophy of Sciences, 1964–1966*. In Memory of Norwood Russell Hanson. [Boston Studies in the Philosophy of Science, Vol. III] 1967 ISBN 90-277-0013-3
15. C. D. Broad, *Induction, Probability, and Causation*. Selected Papers. 1968 ISBN 90-277-0012-5
16. G. Patzig, *Aristotle's Theory of the Syllogism*. A Logical-philosophical Study of *Book A* of the *Prior Analytics*. Translated from German by J. Barnes. 1968 ISBN 90-277-0030-3
17. N. Rescher, *Topics in Philosophical Logic*. 1968 ISBN 90-277-0084-2
18. R. S. Cohen and M. W. Wartofsky (eds.), *Proceedings of the Boston Colloquium for the Philosophy of Science, 1966–1968, Part I*. [Boston Studies in the Philosophy of Science, Vol. IV] 1969 ISBN 90-277-0014-1
19. R. S. Cohen and M. W. Wartofsky (eds.), *Proceedings of the Boston Colloquium for the Philosophy of Science, 1966–1968, Part II*. [Boston Studies in the Philosophy of Science, Vol. V] 1969 ISBN 90-277-0015-X

20. J. W. Davis, D. J. Hockney and W. K. Wilson (eds.), *Philosophical Logic*. 1969
ISBN 90-277-0075-3
21. D. Davidson and J. Hintikka (eds.), *Words and Objections*. Essays on the Work of W. V. Quine. 1969, rev. ed. 1975 ISBN 90-277-0074-5; Pb 90-277-0602-6
22. P. Suppes, *Studies in the Methodology and Foundations of Science. Selected Papers from 1951 to 1969*. 1969 ISBN 90-277-0020-6
23. J. Hintikka, *Models for Modalities*. Selected Essays. 1969
ISBN 90-277-0078-8; Pb 90-277-0598-4
24. N. Rescher *et al.* (eds.), *Essays in Honor of Carl G. Hempel*. A Tribute on the Occasion of His 65th Birthday. 1969 ISBN 90-277-0085-0
25. P. V. Tavanec (ed.), *Problems of the Logic of Scientific Knowledge*. Translated from Russian. 1970 ISBN 90-277-0087-7
26. M. Swain (ed.), *Induction, Acceptance, and Rational Belief*. 1970
ISBN 90-277-0086-9
27. R. S. Cohen and R. J. Seeger (eds.), *Ernst Mach: Physicist and Philosopher*. [Boston Studies in the Philosophy of Science, Vol. VI]. 1970 ISBN 90-277-0016-8
28. J. Hintikka and P. Suppes, *Information and Inference*. 1970 ISBN 90-277-0155-5
29. K. Lambert, *Philosophical Problems in Logic*. Some Recent Developments. 1970
ISBN 90-277-0079-6
30. R. A. Eberle, *Nominalistic Systems*. 1970 ISBN 90-277-0161-X
31. P. Weingartner and G. Zecha (eds.), *Induction, Physics, and Ethics*. 1970
ISBN 90-277-0158-X
32. E. W. Beth, *Aspects of Modern Logic*. Translated from Dutch. 1970
ISBN 90-277-0173-3
33. R. Hilpinen (ed.), *Deontic Logic*. Introductory and Systematic Readings. 1971
See also No. 152. ISBN Pb (1981 rev.) 90-277-1302-2
34. J.-L. Krivine, *Introduction to Axiomatic Set Theory*. Translated from French. 1971
ISBN 90-277-0169-5; Pb 90-277-0411-2
35. J. D. Sneed, *The Logical Structure of Mathematical Physics*. 2nd rev. ed., 1979
ISBN 90-277-1056-2; Pb 90-277-1059-7
36. C. R. Kordig, *The Justification of Scientific Change*. 1971
ISBN 90-277-0181-4; Pb 90-277-0475-9
37. M. Čapek, *Bergson and Modern Physics*. A Reinterpretation and Re-evaluation. [Boston Studies in the Philosophy of Science, Vol. VII] 1971 ISBN 90-277-0186-5
38. N. R. Hanson, *What I Do Not Believe, and Other Essays*. Ed. by S. Toulmin and H. Woolf. 1971 ISBN 90-277-0191-1
39. R. C. Buck and R. S. Cohen (eds.), *PSA 1970*. Proceedings of the Second Biennial Meeting of the Philosophy of Science Association, Boston, Fall 1970. In Memory of Rudolf Carnap. [Boston Studies in the Philosophy of Science, Vol. VIII] 1971
ISBN 90-277-0187-3; Pb 90-277-0309-4
40. D. Davidson and G. Harman (eds.), *Semantics of Natural Language*. 1972
ISBN 90-277-0304-3; Pb 90-277-0310-8
41. Y. Bar-Hillel (ed.), *Pragmatics of Natural Languages*. 1971
ISBN 90-277-0194-6; Pb 90-277-0599-2
42. S. Stenlund, *Combinators, λ-Terms and Proof Theory*. 1972 ISBN 90-277-0305-1
43. M. Strauss, *Modern Physics and Its Philosophy*. Selected Paper in the Logic, History, and Philosophy of Science. 1972 ISBN 90-277-0230-6
44. M. Bunge, *Method, Model and Matter*. 1973 ISBN 90-277-0252-7
45. M. Bunge, *Philosophy of Physics*. 1973 ISBN 90-277-0253-5

46. A. A. Zinov'ev, *Foundations of the Logical Theory of Scientific Knowledge (Complex Logic)*. Revised and enlarged English edition with an appendix by G. A. Smirnov, E. A. Sidorenka, A. M. Fedina and L. A. Bobrova. [Boston Studies in the Philosophy of Science, Vol. IX] 1973 ISBN 90-277-0193-8; Pb 90-277-0324-8
47. L. Tondl, *Scientific Procedures*. A Contribution concerning the Methodological Problems of Scientific Concepts and Scientific Explanation. Translated from Czech by D. Short. Edited by R.S. Cohen and M.W. Wartofsky. [Boston Studies in the Philosophy of Science, Vol. X] 1973 ISBN 90-277-0147-4; Pb 90-277-0323-X
48. N. R. Hanson, *Constellations and Conjectures*. 1973 ISBN 90-277-0192-X
49. K. J. J. Hintikka, J. M. E. Moravcsik and P. Suppes (eds.), *Approaches to Natural Language*. 1973 ISBN 90-277-0220-9; Pb 90-277-0233-0
50. M. Bunge (ed.), *Exact Philosophy*. Problems, Tools and Goals. 1973 ISBN 90-277-0251-9
51. R. J. Bogdan and I. Niiniluoto (eds.), *Logic, Language and Probability*. 1973 ISBN 90-277-0312-4
52. G. Pearce and P. Maynard (eds.), *Conceptual Change*. 1973 ISBN 90-277-0287-X; Pb 90-277-0339-6
53. I. Niiniluoto and R. Tuomela, *Theoretical Concepts and Hypothetico-inductive Inference*. 1973 ISBN 90-277-0343-4
54. R. Fraissé, *Course of Mathematical Logic* – Volume 1: *Relation and Logical Formula*. Translated from French. 1973 ISBN 90-277-0268-3; Pb 90-277-0403-1 (For *Volume 2* see under No. 69).
55. A. Grünbaum, *Philosophical Problems of Space and Time*. Edited by R.S. Cohen and M.W. Wartofsky. 2nd enlarged ed. [Boston Studies in the Philosophy of Science, Vol. XII] 1973 ISBN 90-277-0357-4; Pb 90-277-0358-2
56. P. Suppes (ed.), *Space, Time and Geometry*. 1973 ISBN 90-277-0386-8; Pb 90-277-0442-2
57. H. Kelsen, *Essays in Legal and Moral Philosophy*. Selected and introduced by O. Weinberger. Translated from German by P. Heath. 1973 ISBN 90-277-0388-4
58. R. J. Seeger and R. S. Cohen (eds.), *Philosophical Foundations of Science*. [Boston Studies in the Philosophy of Science, Vol. XI] 1974 ISBN 90-277-0390-6; Pb 90-277-0376-0
59. R. S. Cohen and M. W. Wartofsky (eds.), *Logical and Epistemological Studies in Contemporary Physics*. [Boston Studies in the Philosophy of Science, Vol. XIII] 1973 ISBN 90-277-0391-4; Pb 90-277-0377-9
60. R. S. Cohen and M. W. Wartofsky (eds.), *Methodological and Historical Essays in the Natural and Social Sciences. Proceedings of the Boston Colloquium for the Philosophy of Science, 1969–1972*. [Boston Studies in the Philosophy of Science, Vol. XIV] 1974 ISBN 90-277-0392-2; Pb 90-277-0378-7
61. R. S. Cohen, J. J. Stachel and M. W. Wartofsky (eds.), *For Dirk Struik. Scientific, Historical and Political Essays*. [Boston Studies in the Philosophy of Science, Vol. XV] 1974 ISBN 90-277-0393-0; Pb 90-277-0379-5
62. K. Ajdukiewicz, *Pragmatic Logic*. Translated from Polish by O. Wojtasiewicz. 1974 ISBN 90-277-0326-4
63. S. Stenlund (ed.), *Logical Theory and Semantic Analysis*. Essays dedicated to Stig Kanger on His 50th Birthday. 1974 ISBN 90-277-0438-4
64. K. F. Schaffner and R. S. Cohen (eds.), *PSA 1972. Proceedings of the Third Biennial Meeting of the Philosophy of Science Association*. [Boston Studies in the Philosophy of Science, Vol. XX] 1974 ISBN 90-277-0408-2; Pb 90-277-0409-0

65. H. E. Kyburg, Jr., *The Logical Foundations of Statistical Inference*. 1974
ISBN 90-277-0330-2; Pb 90-277-0430-9
66. M. Grene, *The Understanding of Nature*. Essays in the Philosophy of Biology. [Boston Studies in the Philosophy of Science, Vol. XXIII] 1974
ISBN 90-277-0462-7; Pb 90-277-0463-5
67. J. M. Broekman, *Structuralism: Moscow, Prague, Paris*. Translated from German. 1974 ISBN 90-277-0478-3
68. N. Geschwind, *Selected Papers on Language and the Brain*. [Boston Studies in the Philosophy of Science, Vol. XVI] 1974 ISBN 90-277-0262-4; Pb 90-277-0263-2
69. R. Fraissé, *Course of Mathematical Logic* – Volume 2: *Model Theory*. Translated from French. 1974 ISBN 90-277-0269-1; Pb 90-277-0510-0
(For *Volume 1* see under No. 54)
70. A. Grzegorczyk, *An Outline of Mathematical Logic*. Fundamental Results and Notions explained with all Details. Translated from Polish. 1974
ISBN 90-277-0359-0; Pb 90-277-0447-3
71. F. von Kutschera, *Philosophy of Language*. 1975 ISBN 90-277-0591-7
72. J. Manninen and R. Tuomela (eds.), *Essays on Explanation and Understanding*. Studies in the Foundations of Humanities and Social Sciences. 1976
ISBN 90-277-0592-5
73. J. Hintikka (ed.), *Rudolf Carnap, Logical Empiricist*. Materials and Perspectives. 1975 ISBN 90-277-0583-6
74. M. Čapek (ed.), *The Concepts of Space and Time*. Their Structure and Their Development. [Boston Studies in the Philosophy of Science, Vol. XXII] 1976
ISBN 90-277-0355-8; Pb 90-277-0375-2
75. J. Hintikka and U. Remes, *The Method of Analysis*. Its Geometrical Origin and Its General Significance. [Boston Studies in the Philosophy of Science, Vol. XXV] 1974 ISBN 90-277-0532-1; Pb 90-277-0543-7
76. J. E. Murdoch and E. D. Sylla (eds.), *The Cultural Context of Medieval Learning*. [Boston Studies in the Philosophy of Science, Vol. XXVI] 1975
ISBN 90-277-0560-7; Pb 90-277-0587-9
77. S. Amsterdamski, *Between Experience and Metaphysics*. Philosophical Problems of the Evolution of Science. [Boston Studies in the Philosophy of Science, Vol. XXXV] 1975 ISBN 90-277-0568-2; Pb 90-277-0580-1
78. P. Suppes (ed.), *Logic and Probability in Quantum Mechanics*. 1976
ISBN 90-277-0570-4; Pb 90-277-1200-X
79. H. von Helmholtz: *Epistemological Writings. The Paul Hertz / Moritz Schlick Centenary Edition of 1921 with Notes and Commentary by the Editors*. Newly translated from German by M. F. Lowe. Edited, with an Introduction and Bibliography, by R. S. Cohen and Y. Elkana. [Boston Studies in the Philosophy of Science, Vol. XXXVII] 1975 ISBN 90-277-0290-X; Pb 90-277-0582-8
80. J. Agassi, *Science in Flux*. [Boston Studies in the Philosophy of Science, Vol. XXVIII] 1975 ISBN 90-277-0584-4; Pb 90-277-0612-2
81. S. G. Harding (ed.), *Can Theories Be Refuted?* Essays on the Duhem-Quine Thesis. 1976 ISBN 90-277-0629-8; Pb 90-277-0630-1
82. S. Nowak, *Methodology of Sociological Research*. General Problems. 1977
ISBN 90-277-0486-4
83. J. Piaget, J.-B. Grize, A. Szemińska and V. Bang, *Epistemology and Psychology of Functions*. Translated from French. 1977 ISBN 90-277-0804-5

84. M. Grene and E. Mendelsohn (eds.), *Topics in the Philosophy of Biology.* [Boston Studies in the Philosophy of Science, Vol. XXVII] 1976
ISBN 90-277-0595-X; Pb 90-277-0596-8
85. E. Fischbein, *The Intuitive Sources of Probabilistic Thinking in Children.* 1975
ISBN 90-277-0626-3; Pb 90-277-1190-9
86. E. W. Adams, *The Logic of Conditionals.* An Application of Probability to Deductive Logic. 1975 ISBN 90-277-0631-X
87. M. Przełęcki and R. Wójcicki (eds.), *Twenty-Five Years of Logical Methodology in Poland.* Translated from Polish. 1976 ISBN 90-277-0601-8
88. J. Topolski, *The Methodology of History.* Translated from Polish by O. Wojtasiewicz. 1976 ISBN 90-277-0550-X
89. A. Kasher (ed.), *Language in Focus: Foundations, Methods and Systems.* Essays dedicated to Yehoshua Bar-Hillel. [Boston Studies in the Philosophy of Science, Vol. XLIII] 1976 ISBN 90-277-0644-1; Pb 90-277-0645-X
90. J. Hintikka, *The Intentions of Intentionality and Other New Models for Modalities.* 1975 ISBN 90-277-0633-6; Pb 90-277-0634-4
91. W. Stegmüller, *Collected Papers on Epistemology, Philosophy of Science and History of Philosophy.* 2 Volumes. 1977 Set ISBN 90-277-0767-7
92. D. M. Gabbay, *Investigations in Modal and Tense Logics with Applications to Problems in Philosophy and Linguistics.* 1976 ISBN 90-277-0656-5
93. R. J. Bogdan, *Local Induction.* 1976 ISBN 90-277-0649-2
94. S. Nowak, *Understanding and Prediction.* Essays in the Methodology of Social and Behavioral Theories. 1976 ISBN 90-277-0558-5; Pb 90-277-1199-2
95. P. Mittelstaedt, *Philosophical Problems of Modern Physics.* [Boston Studies in the Philosophy of Science, Vol. XVIII] 1976 ISBN 90-277-0285-3; Pb 90-277-0506-2
96. G. Holton and W. A. Blanpied (eds.), *Science and Its Public: The Changing Relationship.* [Boston Studies in the Philosophy of Science, Vol. XXXIII] 1976
ISBN 90-277-0657-3; Pb 90-277-0658-1
97. M. Brand and D. Walton (eds.), *Action Theory.* 1976 ISBN 90-277-0671-9
98. P. Gochet, *Outline of a Nominalist Theory of Propositions.* An Essay in the Theory of Meaning and in the Philosophy of Logic. 1980 ISBN 90-277-1031-7
99. R. S. Cohen, P. K. Feyerabend, and M. W. Wartofsky (eds.), *Essays in Memory of Imre Lakatos.* [Boston Studies in the Philosophy of Science, Vol. XXXIX] 1976
ISBN 90-277-0654-9; Pb 90-277-0655-7
100. R. S. Cohen and J. J. Stachel (eds.), *Selected Papers of Léon Rosenfield.* [Boston Studies in the Philosophy of Science, Vol. XXI] 1979
ISBN 90-277-0651-4; Pb 90-277-0652-2
101. R. S. Cohen, C. A. Hooker, A. C. Michalos and J. W. van Evra (eds.), *PSA 1974. Proceedings of the 1974 Biennial Meeting of the Philosophy of Science Association.* [Boston Studies in the Philosophy of Science, Vol. XXXII] 1976
ISBN 90-277-0647-6; Pb 90-277-0648-4
102. Y. Fried and J. Agassi, *Paranoia.* A Study in Diagnosis. [Boston Studies in the Philosophy of Science, Vol. L] 1976 ISBN 90-277-0704-9; Pb 90-277-0705-7
103. M. Przełęcki, K. Szaniawski and R. Wójcicki (eds.), *Formal Methods in the Methodology of Empirical Sciences.* 1976 ISBN 90-277-0698-0
104. J. M. Vickers, *Belief and Probability.* 1976 ISBN 90-277-0744-8
105. K. H. Wolff, *Surrender and Catch.* Experience and Inquiry Today. [Boston Studies in the Philosophy of Science, Vol. LI] 1976
ISBN 90-277-0758-8; Pb 90-277-0765-0

106. K. Kosík, *Dialectics of the Concrete*. A Study on Problems of Man and World. [Boston Studies in the Philosophy of Science, Vol. LII] 1976
ISBN 90-277-0761-8; Pb 90-277-0764-2
107. N. Goodman, *The Structure of Appearance*. 3rd ed. with an Introduction by G. Hellman. [Boston Studies in the Philosophy of Science, Vol. LIII] 1977
ISBN 90-277-0773-1; Pb 90-277-0774-X
108. K. Ajdukiewicz, *The Scientific World-Perspective and Other Essays, 1931-1963*. Translated from Polish. Edited and with an Introduction by J. Giedymin. 1978
ISBN 90-277-0527-5
109. R. L. Causey, *Unity of Science*. 1977 ISBN 90-277-0779-0
110. R. E. Grandy, *Advanced Logic for Applications*. 1977 ISBN 90-277-0781-2
111. R. P. McArthur, *Tense Logic*. 1976 ISBN 90-277-0697-2
112. L. Lindahl, *Position and Change*. A Study in Law and Logic. Translated from Swedish by P. Needham. 1977 ISBN 90-277-0787-1
113. R. Tuomela, *Dispositions*. 1978 ISBN 90-277-0810-X
114. H. A. Simon, *Models of Discovery and Other Topics in the Methods of Science*. [Boston Studies in the Philosophy of Science, Vol. LIV] 1977
ISBN 90-277-0812-6; Pb 90-277-0858-4
115. R. D. Rosenkrantz, *Inference, Method and Decision*. Towards a Bayesian Philosophy of Science. 1977 ISBN 90-277-0817-7; Pb 90-277-0818-5
116. R. Tuomela, *Human Action and Its Explanation*. A Study on the Philosophical Foundations of Psychology. 1977 ISBN 90-277-0824-X
117. M. Lazerowitz, *The Language of Philosophy*. Freud and Wittgenstein. [Boston Studies in the Philosophy of Science, Vol. LV] 1977
ISBN 90-277-0826-6; Pb 90-277-0862-2
118. Not published
119. J. Pelc (ed.), *Semiotics in Poland, 1894–1969*. Translated from Polish. 1979
ISBN 90-277-0811-8
120. I. Pörn, *Action Theory and Social Science*. Some Formal Models. 1977
ISBN 90-277-0846-0
121. J. Margolis, *Persons and Mind*. The Prospects of Nonreductive Materialism. [Boston Studies in the Philosophy of Science, Vol. LVII] 1977
ISBN 90-277-0854-1; Pb 90-277-0863-0
122. J. Hintikka, I. Niiniluoto, and E. Saarinen (eds.), *Essays on Mathematical and Philosophical Logic*. 1979 ISBN 90-277-0879-7
123. T. A. F. Kuipers, *Studies in Inductive Probability and Rational Expectation*. 1978
ISBN 90-277-0882-7
124. E. Saarinen, R. Hilpinen, I. Niiniluoto and M. P. Hintikka (eds.), *Essays in Honour of Jaakko Hintikka on the Occasion of His 50th Birthday*. 1979
ISBN 90-277-0916-5
125. G. Radnitzky and G. Andersson (eds.), *Progress and Rationality in Science*. [Boston Studies in the Philosophy of Science, Vol. LVIII] 1978
ISBN 90-277-0921-1; Pb 90-277-0922-X
126. P. Mittelstaedt, *Quantum Logic*. 1978 ISBN 90-277-0925-4
127. K. A. Bowen, *Model Theory for Modal Logic*. Kripke Models for Modal Predicate Calculi. 1979 ISBN 90-277-0929-7
128. H. A. Bursen, *Dismantling the Memorv Machine*. A Philosophical Investigation of Machine Theories of Memory. 1978 ISBN 90-277-0933-5

129. M. W. Wartofsky, *Models*. Representation and the Scientific Understanding. [Boston Studies in the Philosophy of Science, Vol. XLVIII] 1979
ISBN 90-277-0736-7; Pb 90-277-0947-5
130. D. Ihde, *Technics and Praxis*. A Philosophy of Technology. [Boston Studies in the Philosophy of Science, Vol. XXIV] 1979 ISBN 90-277-0953-X; Pb 90-277-0954-8
131. J. J. Wiatr (ed.), *Polish Essays in the Methodology of the Social Sciences*. [Boston Studies in the Philosophy of Science, Vol. XXIX] 1979
ISBN 90-277-0723-5; Pb 90-277-0956-4
132. W. C. Salmon (ed.), *Hans Reichenbach: Logical Empiricist*. 1979
ISBN 90-277-0958-0
133. P. Bieri, R.-P. Horstmann and L. Krüger (eds.), *Transcendental Arguments in Science*. Essays in Epistemology. 1979 ISBN 90-277-0963-7; Pb 90-277-0964-5
134. M. Marković and G. Petrović (eds.), *Praxis*. Yugoslav Essays in the Philosophy and Methodology of the Social Sciences. [Boston Studies in the Philosophy of Science, Vol. XXXVI] 1979 ISBN 90-277-0727-8; Pb 90-277-0968-8
135. R. Wójcicki, *Topics in the Formal Methodology of Empirical Sciences*. Translated from Polish. 1979 ISBN 90-277-1004-X
136. G. Radnitzky and G. Andersson (eds.), *The Structure and Development of Science*. [Boston Studies in the Philosophy of Science, Vol. LIX] 1979
ISBN 90-277-0994-7; Pb 90-277-0995-5
137. J. C. Webb, *Mechanism, Mentalism and Metamathematics*. An Essay on Finitism. 1980 ISBN 90-277-1046-5
138. D. F. Gustafson and B. L. Tapscott (eds.), *Body, Mind and Method*. Essays in Honor of Virgil C. Aldrich. 1979 ISBN 90-277-1013-9
139. L. Nowak, *The Structure of Idealization*. Towards a Systematic Interpretation of the Marxian Idea of Science. 1980 ISBN 90-277-1014-7
140. C. Perelman, *The New Rhetoric and the Humanities*. Essays on Rhetoric and Its Applications. Translated from French and German. With an Introduction by H. Zyskind. 1979 ISBN 90-277-1018-X; Pb 90-277-1019-8
141. W. Rabinowicz, *Universalizability*. A Study in Morals and Metaphysics. 1979
ISBN 90-277-1020-2
142. C. Perelman, *Justice, Law and Argument*. Essays on Moral and Legal Reasoning. Translated from French and German. With an Introduction by H.J. Berman. 1980
ISBN 90-277-1089-9; Pb 90-277-1090-2
143. S. Kanger and S. Öhman (eds.), *Philosophy and Grammar*. Papers on the Occasion of the Quincentennial of Uppsala University. 1981 ISBN 90-277-1091-0
144. T. Pawlowski, *Concept Formation in the Humanities and the Social Sciences*. 1980
ISBN 90-277-1096-1
145. J. Hintikka, D. Gruender and E. Agazzi (eds.), *Theory Change, Ancient Axiomatics and Galileo's Methodology*. Proceedings of the 1978 Pisa Conference on the History and Philosophy of Science, Volume I. 1981 ISBN 90-277-1126-7
146. J. Hintikka, D. Gruender and E. Agazzi (eds.), *Probabilistic Thinking, Thermodynamics, and the Interaction of the History and Philosophy of Science*. Proceedings of the 1978 Pisa Conference on the History and Philosophy of Science, Volume II. 1981 ISBN 90-277-1127-5
147. U. Mönnich (ed.), *Aspects of Philosophical Logic*. Some Logical Forays into Central Notions of Linguistics and Philosophy. 1981 ISBN 90-277-1201-8
148. D. M. Gabbay, *Semantical Investigations in Heyting's Intuitionistic Logic*. 1981
ISBN 90-277-1202-6

149. E. Agazzi (ed.), *Modern Logic – A Survey*. Historical, Philosophical, and Mathematical Aspects of Modern Logic and Its Applications. 1981 ISBN 90-277-1137-2
150. A. F. Parker-Rhodes, *The Theory of Indistinguishables*. A Search for Explanatory Principles below the Level of Physics. 1981 ISBN 90-277-1214-X
151. J. C. Pitt, *Pictures, Images, and Conceptual Change*. An Analysis of Wilfrid Sellars' Philosophy of Science. 1981 ISBN 90-277-1276-X; Pb 90-277-1277-8
152. R. Hilpinen (ed.), *New Studies in Deontic Logic*. Norms, Actions, and the Foundations of Ethics. 1981 ISBN 90-277-1278-6; Pb 90-277-1346-4
153. C. Dilworth, *Scientific Progress*. A Study Concerning the Nature of the Relation between Successive Scientific Theories. 2nd, rev. and augmented ed., 1986 ISBN 90-277-2215-3; Pb 90-277-2216-1
154. D. Woodruff Smith and R. McIntyre, *Husserl and Intentionality*. A Study of Mind, Meaning, and Language. 1982 ISBN 90-277-1392-8; Pb 90-277-1730-3
155. R. J. Nelson, *The Logic of Mind*. 2nd. ed., 1989 ISBN 90-277-2819-4; Pb 90-277-2822-4
156. J. F. A. K. van Benthem, *The Logic of Time*. A Model-Theoretic Investigation into the Varieties of Temporal Ontology, and Temporal Discourse. 1983; 2nd ed., 1991 ISBN 0-7923-1081-0
157. R. Swinburne (ed.), *Space, Time and Causality*. 1983 ISBN 90-277-1437-1
158. E. T. Jaynes, *Papers on Probability, Statistics and Statistical Physics*. Ed. by R. D. Rozenkrantz. 1983 ISBN 90-277-1448-7; Pb (1989) 0-7923-0213-3
159. T. Chapman, *Time: A Philosophical Analysis*. 1982 ISBN 90-277-1465-7
160. E. N. Zalta, *Abstract Objects*. An Introduction to Axiomatic Metaphysics. 1983 ISBN 90-277-1474-6
161. S. Harding and M. B. Hintikka (eds.), *Discovering Reality*. Feminist Perspectives on Epistemology, Metaphysics, Methodology, and Philosophy of Science. 1983 ISBN 90-277-1496-7; Pb 90-277-1538-6
162. M. A. Stewart (ed.), *Law, Morality and Rights*. 1983 ISBN 90-277-1519-X
163. D. Mayr and G. Süssmann (eds.), *Space, Time, and Mechanics*. Basic Structures of a Physical Theory. 1983 ISBN 90-277-1525-4
164. D. Gabbay and F. Guenthner (eds.), *Handbook of Philosophical Logic*. Vol. I: Elements of Classical Logic. 1983 ISBN 90-277-1542-4
165. D. Gabbay and F. Guenthner (eds.), *Handbook of Philosophical Logic*. Vol. II: Extensions of Classical Logic. 1984 ISBN 90-277-1604-8
166. D. Gabbay and F. Guenthner (eds.), *Handbook of Philosophical Logic*. Vol. III: Alternative to Classical Logic. 1986 ISBN 90-277-1605-6
167. D. Gabbay and F. Guenthner (eds.), *Handbook of Philosophical Logic*. Vol. IV: Topics in the Philosophy of Language. 1989 ISBN 90-277-1606-4
168. A. J. I. Jones, *Communication and Meaning*. An Essay in Applied Modal Logic. 1983 ISBN 90-277-1543-2
169. M. Fitting, *Proof Methods for Modal and Intuitionistic Logics*. 1983 ISBN 90-277-1573-4
170. J. Margolis, *Culture and Cultural Entities*. Toward a New Unity of Science. 1984 ISBN 90-277-1574-2
171. R. Tuomela, *A Theory of Social Action*. 1984 ISBN 90-277-1703-6
172. J. J. E. Gracia, E. Rabossi, E. Villanueva and M. Dascal (eds.), *Philosophical Analysis in Latin America*. 1984 ISBN 90-277-1749-4
173. P. Ziff, *Epistemic Analysis*. A Coherence Theory of Knowledge. 1984 ISBN 90-277-1751-7

174. P. Ziff, *Antiaesthetics*. An Appreciation of the Cow with the Subtile Nose. 1984 ISBN 90-277-1773-7
175. W. Balzer, D. A. Pearce, and H.-J. Schmidt (eds.), *Reduction in Science*. Structure, Examples, Philosophical Problems. 1984 ISBN 90-277-1811-3
176. A. Peczenik, L. Lindahl and B. van Roermund (eds.), *Theory of Legal Science*. Proceedings of the Conference on Legal Theory and Philosophy of Science (Lund, Sweden, December 1983). 1984 ISBN 90-277-1834-2
177. I. Niiniluoto, *Is Science Progressive?* 1984 ISBN 90-277-1835-0
178. B. K. Matilal and J. L. Shaw (eds.), *Analytical Philosophy in Comparative Perspective*. Exploratory Essays in Current Theories and Classical Indian Theories of Meaning and Reference. 1985 ISBN 90-277-1870-9
179. P. Kroes, *Time: Its Structure and Role in Physical Theories*. 1985 ISBN 90-277-1894-6
180. J. H. Fetzer, *Sociobiology and Epistemology*. 1985 ISBN 90-277-2005-3; Pb 90-277-2006-1
181. L. Haaparanta and J. Hintikka (eds.), *Frege Synthesized*. Essays on the Philosophical and Foundational Work of Gottlob Frege. 1986 ISBN 90-277-2126-2
182. M. Detlefsen, *Hilbert's Program*. An Essay on Mathematical Instrumentalism. 1986 ISBN 90-277-2151-3
183. J. L. Golden and J. J. Pilotta (eds.), *Practical Reasoning in Human Affairs*. Studies in Honor of Chaim Perelman. 1986 ISBN 90-277-2255-2
184. H. Zandvoort, *Models of Scientific Development and the Case of Nuclear Magnetic Resonance*. 1986 ISBN 90-277-2351-6
185. I. Niiniluoto, *Truthlikeness*. 1987 ISBN 90-277-2354-0
186. W. Balzer, C. U. Moulines and J. D. Sneed, *An Architectonic for Science*. The Structuralist Program. 1987 ISBN 90-277-2403-2
187. D. Pearce, *Roads to Commensurability*. 1987 ISBN 90-277-2414-8
188. L. M. Vaina (ed.), *Matters of Intelligence*. Conceptual Structures in Cognitive Neuroscience. 1987 ISBN 90-277-2460-1
189. H. Siegel, *Relativism Refuted*. A Critique of Contemporary Epistemological Relativism. 1987 ISBN 90-277-2469-5
190. W. Callebaut and R. Pinxten, *Evolutionary Epistemology*. A Multiparadigm Program, with a Complete Evolutionary Epistemology Bibliograph. 1987 ISBN 90-277-2582-9
191. J. Kmita, *Problems in Historical Epistemology*. 1988 ISBN 90-277-2199-8
192. J. H. Fetzer (ed.), *Probability and Causality*. Essays in Honor of Wesley C. Salmon, with an Annotated Bibliography. 1988 ISBN 90-277-2607-8; Pb 1-5560-8052-2
193. A. Donovan, L. Laudan and R. Laudan (eds.), *Scrutinizing Science*. Empirical Studies of Scientific Change. 1988 ISBN 90-277-2608-6
194. H.R. Otto and J.A. Tuedio (eds.), *Perspectives on Mind*. 1988 ISBN 90-277-2640-X
195. D. Batens and J.P. van Bendegem (eds.), *Theory and Experiment*. Recent Insights and New Perspectives on Their Relation. 1988 ISBN 90-277-2645-0
196. J. Österberg, *Self and Others*. A Study of Ethical Egoism. 1988 ISBN 90-277-2648-5
197. D.H. Helman (ed.), *Analogical Reasoning*. Perspectives of Artificial Intelligence, Cognitive Science, and Philosophy. 1988 ISBN 90-277-2711-2
198. J. Wolenski, *Logic and Philosophy in the Lvov-Warsaw School*. 1989 ISBN 90-277-2749-X

199. R. Wójcicki, *Theory of Logical Calculi*. Basic Theory of Consequence Operations. 1988 ISBN 90-277-2785-6
200. J. Hintikka and M.B. Hintikka, *The Logic of Epistemology and the Epistemology of Logic*. Selected Essays. 1989 ISBN 0-7923-0040-8; Pb 0-7923-0041-6
201. E. Agazzi (ed.), *Probability in the Sciences*. 1988 ISBN 90-277-2808-9
202. M. Meyer (ed.), *From Metaphysics to Rhetoric*. 1989 ISBN 90-277-2814-3
203. R.L. Tieszen, *Mathematical Intuition*. Phenomenology and Mathematical Knowledge. 1989 ISBN 0-7923-0131-5
204. A. Melnick, *Space, Time, and Thought in Kant*. 1989 ISBN 0-7923-0135-8
205. D.W. Smith, *The Circle of Acquaintance*. Perception, Consciousness, and Empathy. 1989 ISBN 0-7923-0252-4
206. M.H. Salmon (ed.), *The Philosophy of Logical Mechanism*. Essays in Honor of Arthur W. Burks. With his Responses, and with a Bibliography of Burk's Work. 1990 ISBN 0-7923-0325-3
207. M. Kusch, *Language as Calculus vs. Language as Universal Medium*. A Study in Husserl, Heidegger, and Gadamer. 1989 ISBN 0-7923-0333-4
208. T.C. Meyering, *Historical Roots of Cognitive Science*. The Rise of a Cognitive Theory of Perception from Antiquity to the Nineteenth Century. 1989 ISBN 0-7923-0349-0
209. P. Kosso, *Observability and Observation in Physical Science*. 1989 ISBN 0-7923-0389-X
210. J. Kmita, *Essays on the Theory of Scientific Cognition*. 1990 ISBN 0-7923-0441-1
211. W. Sieg (ed.), *Acting and Reflecting*. The Interdisciplinary Turn in Philosophy. 1990 ISBN 0-7923-0512-4
212. J. Karpiński, *Causality in Sociological Research*. 1990 ISBN 0-7923-0546-9
213. H.A. Lewis (ed.), *Peter Geach: Philosophical Encounters*. 1991 ISBN 0-7923-0823-9
214. M. Ter Hark, *Beyond the Inner and the Outer*. Wittgenstein's Philosophy of Psychology. 1990 ISBN 0-7923-0850-6
215. M. Gosselin, *Nominalism and Contemporary Nominalism*. Ontological and Epistemological Implications of the Work of W.V.O. Quine and of N. Goodman. 1990 ISBN 0-7923-0904-9
216. J.H. Fetzer, D. Shatz and G. Schlesinger (eds.), *Definitions and Definability*. Philosophical Perspectives. 1991 ISBN 0-7923-1046-2
217. E. Agazzi and A. Cordero (eds.), *Philosophy and the Origin and Evolution of the Universe*. 1991 ISBN 0-7923-1322-4
218. M. Kusch, *Foucault's Strata and Fields*. An Investigation into Archaeological and Genealogical Science Studies. 1991 ISBN 0-7923-1462-X

Previous volumes are still available.

KLUWER ACADEMIC PUBLISHERS – DORDRECHT / BOSTON / LONDON

GPSR Compliance

The European Union's (EU) General Product Safety Regulation (GPSR) is a set of rules that requires consumer products to be safe and our obligations to ensure this.

If you have any concerns about our products, you can contact us on ProductSafety@springernature.com

In case Publisher is established outside the EU, the EU authorized representative is:

Springer Nature Customer Service Center GmbH
Europaplatz 3
69115 Heidelberg, Germany

Batch number: 10823811

Printed by Printforce, the Netherlands